AF587846

DNA AND RNA:
PROPERTIES AND MODIFICATIONS, FUNCTIONS AND INTERACTIONS,
RECOMBINATION AND APPLICATIONS

MICRORNA *LET-7*

ROLE IN HUMAN DISEASES AND DRUG DISCOVERY

DNA and RNA: Properties and Modifications, Functions and Interactions, Recombination and Applications

Additional books in this series can be found on Nova's website under the Series tab.

Additional E-books in this series can be found on Nova's website under the E-book tab.

Genetics - Research and Issues

Additional books in this series can be found on Nova's website under the Series tab.

Additional E-books in this series can be found on Nova's website under the E-book tab.

DNA AND RNA:
PROPERTIES AND MODIFICATIONS, FUNCTIONS AND INTERACTIONS,
RECOMBINATION AND APPLICATIONS

MICRORNA *LET-7*

ROLE IN HUMAN DISEASES AND DRUG DISCOVERY

NEETU DAHIYA
EDITOR

Nova Science Publishers, Inc.
New York

For permission to use material from this book please contact us:
Telephone 631-231-7269; Fax 631-231-8175
Web Site: http://www.novapublishers.com

LIBRARY OF CONGRESS CATALOGING-IN-PUBLICATION DATA

Library of Congress Control Number: 2012933150

ISBN 978-1-62081-152-8

Published by Nova Science Publishers, Inc. † New York

Contents

Preface		**vii**
Chapter 1	MicroRNA *Let-7*: Discovery, Biogenesis, and General Function *Cheng-Chia Yu*	**1**
Chapter 2	Transcriptional and Post-Transcriptional Regulation of *Let-7* *Monika Sangwan and Neetu Dahiya*	**11**
Chapter 3	*Let-7* and Signaling Pathways in Cancers *Yuanjia Tang and Nan Shen*	**25**
Chapter 4	Role of *Let-7* Networks in Stem Cell Maintenance and Differentiation *Candida Vaz and Vivek Tanavde*	**43**
Chapter 5	The Role of Microrna *Let-7* in Stem/Progenitor Cells *Shao-Chih Chiu and Shinn-Zong Lin*	**57**
Chapter 6	The Impact of the *Let-7* Family on Resistance to Anticancer Treatment *M. Fritz, D. J. Hussey, J. Haier and R. Hummel*	**75**
Chapter 7	Therapeutic Potential of Microrna *Let-7* in Human Diseases *Elaine Lu Wang, Xu-Hui Li and Zhi Rong Qian*	**89**
Chapter 8	*Let-7* and Cancer *Cesar Seigi Fuziwara, Murilo Vieira Geraldo and Edna Teruko Kimura*	**109**
Chapter 9	Roles of *Let-7* Family Mirnas in Development and Differentiation *Rikako Sanuki and Takahisa Furukawa*	**125**
Index		**145**

PREFACE

MicroRNAs are small non-coding RNAs involved in posttranscriptional regulation of gene expression. Thousands of miRNAs have been identified in different organisms including viruses, insects, plants and animals. MiRNAs has emerged as key regulators of important biological processes. The differential expression of miRNAs in various human diseases has made them potential candidates for developing novel therapies and personalized medicines. This book is focused on microRNA *Let-7*, the second miRNA discovered in the year 2000 and one of the most studied miRNA. This book discusses various aspects of miRNA *Let-7* starting from its discovery, biogenesis, transcriptional and posttranscriptional regulation to its crucial role in various fundamental cellular processes such as development, stem cell maintenance and differentiation, regulation of signaling pathways in cancer, drug resistance and therapeutic potential in different human diseases.

Chapter 1 - MiRNAs are small non coding RNAs capable of regulating gene expression at post-transcriptional level. The discovery of miRNAs has revolutionized the field of biomedical research. A large portion of human genome is regulated by miRNAs. They target a number of genes such as cell cycle regulators, transcription factors, and are involved in many aspects of tumor development, progression and metastasis. Tumor suppressive microRNA *Let-7* is an important miRNA, originally discovered in the *Caenorhabditis elegans*. Several *Let-7* targets have been already identified among which RAS and HMGA2 has oncogenic roles. The tumor suppressive function of *Let-7* has made it an important therapeutic target. Thus, *Let-7* may be a useful tumor marker and a potential candidate gene for miRNA-based therapies for the treatment of cancer. This chapter will discuss about *Let-7* discovery, biogenesis and its common functions.

Chapter 2 - MicroRNAs have emerged as novel molecules involved in post-transcriptional regulation of genes. They play important roles in development, differentiation, cellular metabolism and are implicated in a number of human diseases including cancer. A number of studies have shown miRNA mediated regulation of gene expression. To understand the complex network of miRNA:mRNA function, it is important to understand the molecular players of miRNA biogenesis and post-transcriptional processing. MicroRNA *Let-7* is the first known human miRNAs and is highly conserved in different organisms. Most of the human cancers show down-regulation of *Let-7* family members suggesting a tumor suppressive function of these miRNAs. *Let-7* has been shown to be an important candidate for predicting survival of cancer patients. For developing novel *Let-7* based therapeutic approaches, it is very important to understand the biogenesis and post-transcriptional regulation of this miRNA. This chapter summarizes the recent advances on regulatory molecules controlling transcriptional and post-transcriptional processing of *Let-7*.

Chapter 3 - The *Let-7* family is one of the first identified miRNA families, and is highly conserved across animal species in sequence. The *Let-7* family members have identical seed sequences and may overlap sets of targets. *Let-7* interwines with cellular networks to exert its function. *Let-7* has been identified as cancer repressor or oncogene. In this chapter, the authors focus on discussion of *Let-7* and cancer signaling. Particular *Let-7* family members have been shown to be specifically deregulated in certain cancers. Epigenetic regulation, and processing at transcription and post-transcriptional level are probably involved in *Let-7* deregulation in cancers. Deregulation of *Let-7* may contribute to pathological process by changing the threshold of apoptosis and cell cycle checkpoints in a tissue or stage dependent manner. The *Let-7* family regulates not only a single important pathway protein, but rather coordinate gene expression levels on a pathway-wide scale. *Let-7*-regulated genes were found enriched in focal adhesion, apoptosis, MAPK and cell cycle signaling pathways. Manipulation of *Let-7* expression may provide a potential strategy for designing novel anti-cancer therapies.

Chapter 4 - *Let-7* was identified in a genetic screen in *C. elegans*. *Let-7* regulates a number of fundamental activities in the cells such as cell division, proliferation and differentiation. *Let-7* is involved in a regulatory feed back loop with LIN28, which is one of the factors required for pluripotency of cells suggesting role of *Let-7* in stem cell differentiation and maintenance. In addition to the *Let-7*/LIN28 feedback loop which is important in cancer stem cell, other *Let-7* targets such as RAS and HMGA2 are essential for self renewal of cancer stem cells and maintenance of the undifferentiated state. *Let-7* promotes differentiation by suppressing self renewal pathways in the cells. The ability of *Let-7* members to regulate multiple pathways plays an important in fine tuning of cellular processes of self-renewal and differentiation.

Chapter 5 - One of the first discovered human microRNAs, *Let-7* and its family members, are highly conserved across species in the regulation of embryonic development and stemness. Misregulation of *Let-7* results in a dedifferentiated cellular state and the development of cell-based disorders such as cancer. Although, many preclinical studies have been devoted to the *Let-7*-based cancer therapy, research into what is the physiological function of *Let-7* in the regeneration of adult tissues has only just begun. The exact role of *Let-7* in cancer is not yet fully understood. There is a need to understand the biological meanings of *Let-7* alterations in normal stem cells and cancers before proceeding to clinical applications. This chapter highlights an outlook on certain pivotal aspects of *Let-7* in stemness and tumorigenesis for future therapeutic potential.

Chapter 6 - MiRNAs (miRNAs) represent a class of naturally occurring, small (19 to 25-nucleotides), non-coding RNA molecules. Among 16000 miRNA genes which have been identified in various species, more than 1,000 miRNAs have been reported in the humans. The lethal-7 (*Let-7*) gene family, which was initially found as an essential developmental gene family in *Caenorhabditis elegans,* is one of the most extensively studied miRNA families, and there is high evidence that these miRNAs play key roles in the initiation, growth and progression of a variety of (malignant) tumors. Based on these data, there is now a growing interest in using these molecules for clinical purposes. The present chapter therefore discusses and highlights potential diagnostic applications of members of the *Let-7* family as predictors of clinical outcome and as markers for response monitoring during chemo- and/ or radiotherapy. In addition, the authors present first promising data on the *Let-7* family concerning their use as primary anticancer agents and, perhaps more importantly, as modifiers and enhancers of well-established anticancer therapy strategies. In this context, modulation of *Let-7* expression seems to provide a new and very promising first-line treatment option in cancer, as well as an additive therapeutic modality to chemo- and radiotherapy, augmenting their effects and possibly reversing drug resistance – a major obstacle in modern anticancer treatment.

Chapter 7 - MicroRNAs (MiRNAs) are emerging as important therapeutic molecules. Their role in regulating fundamental processes of life and altered expression in a number of human diseases has made it necessary to study their diagnostic and therapeutic potential. A number of human cancers shows unique miRNA signature useful in distinguishing different subtypes of tumors. MiRNAs has been reported to function both as oncogenes and tumor suppressors. For example members of *Let-7* family are well known tumor suppressors whereas miR-17-92 family members function as oncogenes. Most of the *Let-7* family members are downregulated in cancers and two well know oncogenic targets regulated by *Let-7* are RAS and HMGA2. *Let-7* is also involved in development of drug resistance. In addition to cancer altered expression of *Let-7* has been reported in other diseases such as lung fibrosis, inflammation etc. Because of its involvement in development and progression of the diseases, *Let-7* is being studied for its potential as a therapeutic target. This chapter summarizes the current knowledge of *Let-7* functions in cancer and other disease conditions and challenges associated with development of miRNA based therapeutic approaches.

Chapter 8 - In cancer, microRNAs dysregulation affects the post-transcriptional control of oncogenes and tumor suppressors. *Let-7* importance emerged in cancer when its implication in regulating the expression of RAS oncogene family was uncovered. Frequently reduced in several types of cancer, *Let-7* exerts a tumor suppressor role in tumorigenesis, and loss of its expression is associated with poor prognosis. The molecular mechanisms of *Let-7* down-regulation involve genetic alterations such as deletions and loss of heterozygosity, and more recently, post-transcriptional regulation by LIN28, which impairs *Let-7* miRNA maturation. The biological role of *Let-7* is being elucidated gradually by the validation of its targets in relevant pathways such as cyclins, CDKs, E2F2 and MYC in cell cycle; RAS in MAPK pathway; IGF2BP1 in drug resistance; FAS and CASP3 in apoptotic process. As the functional understanding of *Let-7* targets improves, new insights into *Let-7* regulatory network and its potential implication in therapeutics arise. Restoring *Let-7* levels *in vitro* leads to growth restrain and apoptosis of cancer cells, suggesting that in the future therapeutical replacement of *Let-7* could improve or potentiate cancer patient response to common treatment. Thus, unraveling molecular *Let-7* network in cancer is essential to address novel perspectives to translational medicine.

Chapter 9 - MiRNAs (miRNAs) are a class of 18 to 25-nucleotide, short, non-coding RNAs that regulate gene expression post-transcriptionally. The lethal-7 (*Let-7*) gene is one of the founding members of the first two miRNA families identified in *C. elegans*, and the first miRNA known to be found in humans. *Let-7* family miRNAs are expressed from early development through maturity in both invertebrates and vertebrates. The majority of miRNAs are not essential for viability or development in *C. elegans*. However, *Let-7* family miRNAs are critical regulators of development that control both early and late developmental timing decisions, and were elucidated by extensive genetic and molecular analysis using *C. elegans* mutants as well as *Drosophila* mutants. In contrast to invertebrates, the authors' understanding of the functional roles of *Let-7* family miRNAs in vertebrate development is still limited because there are many redundant *Let-7* family miRNAs, and thus experiments to clarify *in vivo* functions of *Let-7* family miRNAs using *Let-7* family miRNA mutants in vertebrates are practically difficult. However, indirect but significant *in vivo* evidence has suggested that *Let-7* is essential for gene regulation in mouse development. In addition, ectopic expression of *Let-7* in zebrafish embryos leads to early down-regulation of endogenous *Let-7* targets, resulting in severe embryonic developmental abnormalities, and suggesting that *Let-7* plays a significant role in zebrafish development. Although studies on vertebrate *Let-7* family miRNAs are still in the early stages, these studies have provided us with evidence of important biological roles for *Let-7*. Based on past and recent findings, the authors review the expression patterns, biological roles, and functional mechanisms of *Let-7* family miRNAs in normal development of both invertebrates and vertebrates. In addition, the authors describe *in vitro* analysis on the molecular roles of *Let-7* family miRNAs underlying cellular differentiation. Future studies on *Let-7* family miRNAs will provide us with not only important clues to understanding cell differentiation in development, but also useful tools for diagnosis and/or therapies in the medical field.

In: MicroRNA *Let-7*
Editor: Neetu Dahiya

ISBN: 978-1-62081-152-8

Chapter 1

MICRORNA *LET-7*: DISCOVERY, BIOGENESIS, AND GENERAL FUNCTION

Cheng-Chia Yu*
Institute of Oral Science, School of Dentistry, College of Oral Medicine, Chung Shan Medical University, Taiwan

I. ABSTRACT

MiRNAs are small non coding RNAs capable of regulating gene expression at post-transcriptional level. The discovery of miRNAs has revolutionized the field of biomedical research. A large portion of human genome is regulated by miRNAs. They target a number of genes such as cell cycle regulators, transcription factors, and are involved in many aspects of tumor development, progression and metastasis. Tumor suppressive microRNA *Let-7* is an important miRNA, originally discovered in the *Caenorhabditis elegans*. Several *Let-7* targets have been already identified among which RAS and HMGA2 has oncogenic roles. The tumor suppressive function of *Let-7* has made it an important therapeutic target. Thus, *Let-7* may be a useful tumor marker and a potential candidate gene for miRNA-based therapies for the treatment of cancer. This chapter will discuss about *Let-7* discovery, biogenesis and its common functions.

II. INTRODUCTION

MicroRNAs (miRNAs) are a large group of short (18–25 nucleotide long), noncoding, RNA molecules transcribed by RNA polymerase II, or in some cases, by RNA polymerase III (Lewis et al., 2005). Most microRNA-mediated regulation occurs at the post-translational level, primarily through its near-perfect or partial antisense complementary fit against the coding region or 3′untranslated region (UTR) of target mRNA, leading to translational repression and/or degradation (Lewis et al., 2005). Since its discovery in 1993, the biological

* Address for correspondence: Cheng-Chia Yu, Ph.D., Institute of Oral Science, School of Dentistry, College of Oral Medicine, Chung Shan Medical University. No.110, Sec.1, Jianguo N.Rd., Taichung 40201, Taiwan. E-mail: *ccyu@csmu.edu.tw*.

function and biogenesis of miRNAs became popular topics for biomedical researchers. MicroRNAs are emerging as important regulators of cell proliferation, aging, differentiation, stress responses, apoptosis, senescence, pluripotency, and tumorigenesis (Lu et al., 2005; Leung and Sharp 2010). MiRNAs do not only act as oncogenes but also as tumor suppressors, which further implicates their role as therapeutic targets (Croce 2009).

III. The Discovery of *Let-7* Family

Let-7 is the second miRNA discovered in the *Caenorhabditis elegans* and first known human miRNA (Reinhart et al., 2000). The authors performed a genetic screen to identify gene regulating the development timing in *C. elegans*. The genes were screened based on the mutations that suppress the synthetic sterile phenotype of a strain bearing LIN14 and AGL35 mutations. Out of 36 suppressor mutations only one mutation n2853 (mapped to *Let-7*) caused the strongest retarded phenotype and a temperature sensitive lethal phenotype in LIN14 background. Higher expression of *Let-7* resulted in terminal differentiation of cells where as lower expression showed opposite effect suggesting role of *Let-7* as a temporal switch between larval and adult fates.

IV. Biogenesis of MicroRNAs *Let-7*

MiRNAs are transcribed as a primary transcript by RNA polymerase II. Primary miRNA is processed into precursor miRNA (pre-miRNA) by Drosha and DRG8 and is then exported from nucleus to cytoplasm through exportin 5. The pre-miRNA is further processed by Dicer into the mature miRNA (Han et al., 2004). There are 11 *Let-7* members in humans (*Let-7*a-1: 9q22.32; *Let-7*a-2: 11q24.1; *Let-7*a-3: 22q13.31; *Let-7*b: 22q13.31; *Let-7*c: 21q21.1; *Let-7*d: 9q22.32; *Let-7*e: 19q13.41; *Let-7*f-1: 9q22.32; *Let-7*f-2: Xp11.22; *Let-7*g: 3p21.1; *Let-7*i: 12q14.1). *Let-7* biogenesis is regulated at both transcriptional and posttranscriptional level. The transcription factors DAF12, CREBα and HBL1 are known to regulate *Let-7* transcription. At posttranscriptional level, LIN28 is the major regulator of *Let-7* expression which control *Let-7* biogenesis at Dicer as well as Drosha level. The detailed roles of various factors regulating *Let-7* biogenesis are described in the chapter 2 (Sangwan and Dahiya).

V. Functions

Most miRNA-mediated regulation occurs at the posttranscriptional level, primarily through its near-perfect or partial complementary fit against the coding region or 3' untranslated region (UTR) of target mRNA, leading to translational repression and/or mRNA degradation. In rare cases, they may also promote translation (Vasudevan et al., 2007).

1) *Let-7* in Regulating Self-Renewal and Differentiation of Stem Cells

Stem cells or progenitor cells among tissues play an important role in replacing damaged cells. As organisms age, the properties and number of stem cells gradually decline, finally resulting in tissue or organ dysfunction. MiRNAs, which regulate stem cell properties or maintain the ability of cell-renewal are, therefore, crucial in the aging process. In *C. elegans*, nuclear receptor DAF12 and its steroidal ligand directly activate promoters of *Let-7* miRNA family members (Bethke et al., 2009). *Let-7* downregulates hbl-1, which drives progression of epidermal stem cells from second to third larval stage patterns of cell division (Bethke et al., 2009). Upon differentiation process of human and mouse embryonic stem cells, the level of *Let-7* increases (Thomson et al., 2004). Mesenchymal stem cells belong to the adult stem cell category. *Let-7*d regulates senescence of human cord blood-derived multipotent stem cells by regulating HMGA2 and p16. These cells quickly go into senescence within 7–12 passages after isolation (Wagner et al., 2008). miRNA expression profiling has revealed the up-regulation of several miRNAs including hsa-*Let-7*f upon *in vitro* propagation. Genes that regulate stem cell properties, so called stemness genes, govern cell renewal, proliferation, quiescence, or protection from stress. Among them, high mobility group A2 (HMGA2) is capable of modifying chromatin structure and is also implicated in regulating stem cell pluripotency. *Let-7* can reduce HMGA2 expression and increases p16 and p19 expression, governing stem cell aging (Nishino et al., 2010; Tzatsos and Bardeesy, 2008). However, that this happens in young, but not old, mice may suggest that some unknown underlying mechanisms remain to be clarified. One of the characteristics of miRNAs is their huge number of targets, which allows them to regulate genes with diverse functions at the same time. Alterations of miRNA levels in stem cells undergoing differentiation, and miRNA's ability to induce pluripotency, may further suggest their role in regulating stem cell properties.

2) Tumor Suppressive Role of *Let-7* Family in Cancer

The roles of miRNAs in cancers have been extensively investigated in the past few years (Croce, 2004). The relevance of miRNAs in cancer is suggested by the changes in expression patterns and recurrent amplification as well as deletion of miRNA genes in cancers (Croce, 2004). Calin et al. (2002) are the first to report the signature of 13 microRNAs capable of distinguishing between indolent and aggressive chronic lymphocytic leukemia (CLL). Since then, several miRNAs have emerged as candidate component of oncogenes and tumor suppressor networks depending on its main target genes. *Let-7* family members have been described as being down-regulated during cancer progression in various human cancers including lung, gastric, ovarian, colon cancer, leiomyoma and melanoma, renal cancer, liver cancer, and head and neck cancer. *Let-7* has been also shown to act as tumor suppressor, most likely through targeting a number of genes with oncogenic activity such as RAS or high mobility group HMGA2. Comparing to normal human laryngeal cells or nondifferentiated cells, *Let-7*a miRNA expression level is highly suppressed (Long et al., 2009). Furthermore, in the same study, *Let-7*a affected RAS and c-MYC expression in protein level, thus mediated cell apoptotic genes and oncogenes expression (Long et al., 2009). The property of resistance to irradiation or chemotherapy treatment is the major clinical problem of malignant cancer.

Increasing the levels of *Let-7*b sensitize cancer cells to radiation treatment. *Let-7*c is down-regulated in lung cancer lymphoma cells (Leucci et al., 2008). *Let-7*d, a member of the *Let-7* family of miRNAs, has also been shown to act as tumor suppressor, also most likely through targeting RAS or HMGA2 (Yu et al., 2011). The region upstream of *Let-7*d promoter is modulated by transforming growth factor (TGF)-β transcription factor SMAD3 (Pandit et al., 2010). The expression of *Let-7*d is significantly down-regulated in chemo-resistant pancreatic cancer (Li et al., 2009). Dysregulation of *Let-7*d also indicated its possible involvement in the development and progression of ovarian cancer, glioblastoma, lung cancer, colon cancer, and leukemia (Shell et al., 2007; Shao et al., 2011; Garzon et al., 2007). *Let-7*f represses metastasis capability of gastric cancer via targeting MYH9 (Pasquinelli et al., 2000). Expression of *Let-7*e and *Let-7*f is also decreased in prostate and ovarian cancer (Dahiya et al., 2008; Ozen et al., 2008). *Let-7*g is downregulated in pigmented nodular adrenocortical disease via regulating Wnt pathway activity (Iliopoulos et al., 2009). *Let-7*i, a member of the *Let-7* family of microRNAs, targets Toll-like receptor 4 mRNA and limits the expression of this pathogen molecular pattern receptor resulting in decreased primary and mature *Let-7*i expression (Chen et al., 2007) The *Let-7*i expression is approved to be modulated by NFκB p50-C/EBPβ silence complex which induced by microbial insult (O'Hara et al., 2010). *Let-7*i has been approved to regulate cancer proliferation and chemotherapy resistance through targeting progesterone receptor membrane component 1 (PGRMC1), and E2F diffuse large B cell lymphoma (Blower et al., 2008; Bueno et al., 2010). Overexpression of *Let-7*i in lung cancer cells reduced chemoresistance (Blower et al., 2008). Yang and his team discovered that *Let-7*i might be used as a therapeutic target to modulate platinum-based chemotherapy and as a biomarker to predict chemotherapy response and survival of ovarian cancer (Yang et al., 2008). Scapoli et al. (2010) found that underexpression of *Let-7*i is associated with progression to metastatic oral squamous cell carcinomas through microarray analysis. These results suggested that the expression of *Let-7* is responsive to pathogen recognition and regulated through primary-microRNA transcription.

3) Emerging Roles of *Let-7* Family in Cancer Initiating Stem Cells

Recent data have demonstrated that tumors contain a small subpopulation of cells called tumor-initiating cells (TICs) or cancer stem cells (CSCs), which exhibit self-renewing capacities and are responsible for tumor maintenance and metastasis (Rosen and Jordan 2009). With the increasing awareness of the importance of miRNAs in tumorigenicity, accumulating evidence has been reported to state the involvement of miRNAs in CSC-like properties. For example, overexpressed *Let-7*a impairs the self-renewal and tumorigenicity properties of CSC isolated from breast by directly targeting HMGA2 (Yu et al., 2007). Our group first demonstrates that *Let-7*a repressed stemness, tumorigenicity, and chemoresistance property in oral cancer-derived ALDH1 positive cells (Yu et al., 2011). *Let-7* plays as a switch regulating lung CSC-like properties, in which it negatively regulated tumor-initiation, EMT, and drug-resistance through inhibiting stemness genes signature (Yu et al., 2011). Further investigation on the interplay between *Let-7*-mediated mechanisms is urgently needed for therapeutic advantages.

4) Role of *Let-7* in Development and Differentiation

Let-7 regulates development timing in *C. elegans* (Reinhart et al., 2000). *Let-7* is expressed at later stages of development and control transition from larval to adult stages. This role of *Let-7* in regulation of development timing is conserved in Drosophila where ecdysone, the major insect molting hormone modulates *Let-7* expression which first appears before puparium formation and reaches highest levels during pupal development (Sempere et al., 2002). *Let-7* microRNA plays an important role in lens regeneration. In adult newt the lens regeneration involves transdifferentiation, first the pigment epithelial cells of the dorsal iris de-differentiate and then differentiate to the lens cell after lentectony (Nakamura et al., 2010). *Let-7* family is found to be down-regulated during the process of dedifferentiation in dorsal iris (Tsonis et al., 2007). Specifically, up-regulation of *Let-7*b would affect proliferation of both dorsal and ventral pigment epithelial cells (Tsonis et al., 2007). *Let-7* microRNA family is highly conserved across animal species in sequence and function. In nematode maturation, *Let-7* induced cellular differentiation, but alteration of its expression with aging possible serving as a regulator of longevity (Chen et al., 2007). During the aging process of murine neural stem cells, significant increased *Let-7* expression was associated with poor self-renewal abilities (Nishino et al., 2008). In human tissues, higher levels of *Let-7* microRNA expression were also detected at uterine senescent fibroids and aged skeletal muscles (Drummond et al., 2011). It proposed that *Let-7* expression might be an indicator of impaired cell cycle function possible contributing to reduced cellular renewal in aged human tissues (Karali et al., 2010). In mammalian eyes, several microRNAs were identified to have their expressions in the lens, and *Let-7* microRNAs were evident to have its expression profile in the adult murine lens.

VI. Future Perspectives

Since one miRNA can regulate the expression of multiple mRNAs, the possibility of a high frequency of overlap between two miRNAs may also erroneously lead to unexpected results when we try to understand the functions of one miRNA. Traditionally, the major theory of cancer is considered as dysregulation of protein-coding tumor suppressor genes and oncogenes. To date, the discovery of epigenetic regulation provides new explanations and reveals a more complicated network of cancer formation. MiRNAs, a small group of noncoding RNAs, draw more attention than ever and are thought to be a new category of regulatory molecules acting as tumor suppressors or oncogenes. Accumulating knowledge in miRNA brings new perspectives in understanding of cell transformation and tumorigenicity. Further studies are needed to understand the mechanism of dysregulation of miRNA as well as their targets in human disease.

Conclusion

Let-7 family miRNAs regulate tumor-initiating and stemness properties in the cells. Elevated expression of *Let-7* family members decreases cancer stem-like characteristics and makes the cancer cells sensitive to radiation New approaches intending to tackle the microRNAs-mediated cancer progression should be encouraged. Delivery of miRNAs has recently gained much attention as it provides a novel strategy for targeting the cancer cells. Adenovirus vectors-mediated delivery or injection of miRNAs constructed with locked nucleic acid (LNA)- or phosphothiolatediester-modified backbones were previously shown a promising suppressive effect on tumor progression. *Let-7* family may be a paradigm of novel miRNA-based therapeutic target for treatment of human cancers.

References

Bethke A, Fielenbach N, Wang Z, Mangelsdorf DJ, Antebi A. Nuclear hormone receptor regulation of microRNAs controls developmental progression. *Science*.2009;324:95-98.

Blower PE, Chung JH, Verducci JS, Lin S, Park JK, Dai Z, Liu CG, Schmittgen TD, Reinhold WC, Croce CM, Weinstein JN, Sadee W. MicroRNAs modulate the chemosensitivity of tumor cells. *Mol. Cancer Ther*. 2008;7:1-9.

Bueno MJ, Gomez de Cedron M, Laresgoiti U, Fernandez-Piqueras J, Zubiaga AM, Malumbres M. Multiple E2F-induced microRNAs prevent replicative stress in response to mitogenic signaling. *Mol. Cell Biol*. 2010;30:2983-2995.

Calin GA, Dumitru CD, Shimizu M, Bichi R, Zupo S, Noch E, Aldler H, Rattan S, Keating M, Rai K, Rassenti L, Kipps T, Negrini M, Bullrich F, Croce CM. Frequent deletions and down-regulation of micro- RNA genes miR15 and miR16 at 13q14 in chronic lymphocytic leukemia. *Proc. Natl. Acad. Sci. U S A*.2002;99:15524-15529.

Chen XM, Splinter PL, O'Hara SP, LaRusso NF. A cellular micro-RNA, *Let-7*i, regulates Toll-like receptor 4 expression and contributes to cholangiocyte immune responses against Cryptosporidium parvum infection. *J. Biol. Chem.* 2007;282: 28929-28938.

Croce CM. Causes and consequences of microRNA dysregulation in cancer. *Nat. Rev. Genet.* 2009;10:704-714.

Dahiya N, Sherman-Baust CA, Wang TL, Davidson B, Shih Ie M, Zhang Y, Wood W, 3rd, Becker KG, Morin PJ. MicroRNA expression and identification of putative miRNA targets in ovarian cancer. *PLoS One*. 2008;3:e2436.

Drummond MJ, McCarthy JJ, Sinha M, Spratt HM, Volpi E, Esser KA, Rasmussen BB. Aging and microRNA expression in human skeletal muscle: a microarray and bioinformatics analysis. *Physiol. Genomics*.2011;43:595-603.

Garzon R, Pichiorri F, Palumbo T, Visentini M, Aqeilan R, Cimmino A, Wang H, Sun H, Volinia S, Alder H, Calin GA, Liu CG, Andreeff M, Croce CM. MicroRNA gene expression during retinoic acid-induced differentiation of human acute promyelocytic leukemia. *Oncogene*.2007;26:4148-4157.

Han J, Lee Y, Yeom KH, Kim YK, Jin H, Kim VN. The Drosha-DGCR8 complex in primary microRNA processing. *Genes. Dev*. 2004;18:3016-30Pasquinelli.

Hui AB, Lenarduzzi M, Krushel T, Waldron L, Pintilie M, Shi W, Perez-Ordonez B, Jurisica I, O'Sullivan B, Waldron J, Gullane P, Cummings B, Liu FF. Comprehensive MicroRNA profiling for head and neck squamous cell carcinomas. *Clin. Cancer Res*. 2010;16:1129-1139.

Iliopoulos D, Bimpaki EI, Nesterova M, Stratakis CA. MicroRNA signature of primary pigmented nodular adrenocortical disease: clinical correlations and regulation of Wnt signaling. *Cancer Res.* 2009;69:3Pasquinelli8-3282.

Karali M, Peluso I, Gennarino VA, Bilio M, Verde R, Lago G, Dolle P, Banfi S. miRNeye: a microRNA expression atlas of the mouse eye. *BMC Genomics*. 2010;11:715.

Leucci E, Cocco M, Onnis A, De Falco G, van Cleef P, Bellan C, van Rijk A, Nyagol J, Byakika B, Lazzi S, Tosi P, van Krieken H, Leoncini L. MYC translocation-negative classical Burkitt lymphoma cases: an alternative pathogenetic mechanism involving miRNA deregulation. *J. Pathol.*2008;216:440-450.

Leung AK, Sharp PA. MicroRNA functions in stress responses. *Mol. Cell*; 2010;40:205-215.

Lewis BP, Burge CB, Bartel DP. Conserved seed pairing, often flanked by adenosines, indicates that thousands of human genes are microRNA targets. *Cell.* 2005;120:15-20.

Li Y, VandenBoom TG, 2nd, Kong D, Wang Z, Ali S, Philip PA, Sarkar FH. Up-regulation of miR-200 and *Let-7* by natural agents leads to the reversal of epithelial-to-mesenchymal transition in gemcitabine-resistant pancreatic cancer cells. *Cancer Res.*2009;69:6704-6712.

Liang S, He L, Zhao X, Miao Y, Gu Y, Guo C, Xue Z, Dou W, Hu F, Wu K, Nie Y, Fan D. MicroRNA *Let-7*f Inhibits Tumor Invasion and Metastasis by Targeting MYH9 in Human Gastric Cancer. *PLoS One*.2011;6:e18409.

Lin YC, Hsieh LC, Kuo MW, Yu J, Kuo HH, Lo WL, Lin RJ, Yu AL, Li WH. Human TRIM71 and its nematode homologue are targets of *Let-7* microRNA and its zebrafish orthologue is essential for development. *Mol. Biol. Evol.* 2007;24:2525-2534.

Long XB, Sun GB, Hu S, Liang GT, Wang N, Zhang XH, Cao PP, Zhen HT, Cui YH, Liu Z. *Let-7*a microRNA functions as a potential tumor suppressor in human laryngeal cancer. *Oncol. Rep.* 2009;22:1189-1195.

Lu J, Getz G, Miska EA, Alvarez-Saavedra E, Lamb J, Peck D, Sweet-Cordero A, Ebert BL, Mak RH, Ferrando AA, Downing JR, Jacks T, Horvitz HR, Golub TR. MicroRNA expression profiles classify human cancers. *Nature* 2005;435:834-838.

Nakamura K, Maki N, Trinh A, Trask HW, Gui J, Tomlinson CR, Tsonis PA. miRNAs in newt lens regeneration: specific control of proliferation and evidence for miRNA networking. *PLoS One*.2010;5:e12058.

Nishino J, Kim I, Chada K, Morrison SJ. Hmga2 promotes neural stem cell self-renewal in young but not old mice by reducing p16Ink4a and p19Arf Expression. *Cell.* 2008;135:2Pasquinelli-239.

O'Hara SP, Splinter PL, Gajdos GB, Trussoni CE, Fernandez-Zapico ME, Chen XM, LaRusso NF. NFkappaB p50-CCAAT/enhancer-binding protein beta (C/EBPbeta)-mediated transcriptional repression of microRNA *Let-7*i following microbial infection. *J. Biol. Chem.* 2010;285:216-225.

Ozen M, Creighton CJ, Ozdemir M, Ittmann M. Widespread deregulation of microRNA expression in human prostate cancer. *Oncogene*. 2008;Pasquinelli:1788-1793.

Pandit KV, Corcoran D, Yousef H, Yarlagadda M, Tzouvelekis A, Gibson KF, Konishi K, Yousem SA, Singh M, Handley D, Richards T, Selman M, Watkins SC, Pardo A, Ben-Yehudah A, Bouros D, Eickelberg O, Ray P, Benos PV, Kaminski N. Inhibition and role of *Let-7*d in idiopathic pulmonary fibrosis. *Am. J. Respir. Crit. Care Med.* 2010;182:220-229.

Pasquinelli AE, Reinhart BJ, Slack F, Martindale MQ, Kuroda MI, Maller B, Hayward DC, Ball EE, Degnan B, Muller P, Spring J, Srinivasan A, Fishman M, Finnerty J, Corbo J, Levine M, Leahy P, Davidson E, Ruvkun G. Conservation of the sequence and temporal expression of *Let-7* heterochronic regulatory RNA. *Nature*. 2000;408: 86-89.

Reinhart BJ, Slack FJ, Basson M, Pasquinelli AE, Bettinger JC, Rougvie AE, Horvitz HR, Ruvkun G. The 21-nucleotide *Let-7* RNA regulates developmental timing in Caenorhabditis elegans. *Nature*. 2000;403: 901-906.

Rosen JM, Jordan CT. The increasing complexity of the cancer stem cell paradigm. *Science*. 2009;324:1670-1673.

Scapoli L, Palmieri A, Lo Muzio L, Pezzetti F, Rubini C, Girardi A, Farinella F, Mazzotta M, Carinci F. MicroRNA expression profiling of oral carcinoma identifies new markers of tumor progression. *Int. J. Immunopathol. Pharmacol*. 2010;23:
1229-1234.

Sempere LF, Dubrovsky EB, Dubrovskaya VA, Berger EM, Ambros V. The expression of the *Let-7* small regulatory RNA is controlled by ecdysone during metamorphosis in Drosophila melanogaster. *Dev. Biol.* 2002;244:170-179.

Shao M, Rossi S, Chelladurai B, Shimizu M, Ntukogu O, Ivan M, Calin GA, Matei D. PDGF induced microRNA alterations in cancer cells. *Nucleic Acids Res.* 2011;39: 4035-4047.

Shell S, Park SM, Radjabi AR, Schickel R, Kistner EO, Jewell DA, Feig C, Lengyel E, Peter ME. *Let-7* expression defines two differentiation stages of cancer. *Proc. Natl. Acad. Sci. U S A*. 2007;104:11400-11405.

Thomson JM, Parker J, Perou CM, Hammond SM. A custom microarray platform for analysis of microRNA gene expression. *Nat. Methods*. 2004;1:47-53.

Tsonis PA, Call MK, Grogg MW, Sartor MA, Taylor RR, Forge A, Fyffe R, Goldenberg R, Cowper-Sal-lari R, Tomlinson CR. MicroRNAs and regeneration: *Let-7* members as potential regulators of dedifferentiation in lens and inner ear hair cell regeneration of the adult newt. *Biochem. Biophy. Res. Commun*. 2007;362:940-945.

Tzatsos A, Bardeesy N. Ink4a/Arf regulation by *Let-7*b and Hmga2: a genetic pathway governing stem cell aging. *Cell Stem. Cell*.2008;3: 469-470.

Vasudevan S, Tong Y, Steitz JA. Switching from repression to activation: microRNAs can up-regulate translation. *Science*.2007;318:1931-1934.

Wagner W, Horn P, Castoldi M, Diehlmann A, Bork S, Saffrich R, Benes V, Blake J, Pfister S, Eckstein V, Ho AD. Replicative senescence of mesenchymal stem cells: a continuous and organized process. *PLoS One*. 2008;3:e2213.

Yang N, Kaur S, Volinia S, Greshock J, Lassus H, Hasegawa K, Liang S, Leminen A, Deng S, Smith L, Johnstone CN, Chen XM, Liu CG, Huang Q, Katsaros D, Calin GA, Weber BL, Butzow R, Croce CM, Coukos G, Zhang L. MicroRNA microarray identifies *Let-7*i as a novel biomarker and therapeutic target in human epithelial ovarian cancer. *Cancer Res*. 2008;68:10307-10314.

Yu CC, Chen YW, Chiou GY, Tsai LL, Huang PI, Chang CY, Tseng LM, Chiou SH, Yen SH, Chou MY, Chu PY, Lo WL. MicroRNA *Let-7*a represses chemoresistance and tumourigenicity in head and neck cancer via stem-like properties ablation. *Oral Oncol*.2011;47:202-210.

Yu F, Yao H, Zhu P, Zhang X, Pan Q, Gong C, Huang Y, Hu X, Su F, Lieberman J, Song E. *Let-7* regulates self renewal and tumorigenicity of breast cancer cells. *Cell*. 2007;131:1109-1123.

Yu ML, Wang JF, Wang GK, You XH, Zhao XX, Jing Q, Qin YW. Vascular smooth muscle cell proliferation is influenced by *Let-7*d microRNA and its interaction with KRAS. *Circ. J.*2011;75:703-709.

In: MicroRNA *Let-7*
Editor: Neetu Dahiya

ISBN: 978-1-62081-152-8

Chapter 2

TRANSCRIPTIONAL AND POST-TRANSCRIPTIONAL REGULATION OF *LET-7*

Monika Sangwan[1] and Neetu Dahiya[2*]
[1]University of Toronto, Genetics and Development Division, Toronto, Canada
[2]Department of Obstetrics and Gynecology, University of Virginia, Charlottesville, US

1. ABSTRACT

MicroRNAs have emerged as novel molecules involved in post-transcriptional regulation of genes. They play important roles in development, differentiation, cellular metabolism and are implicated in a number of human diseases including cancer. A number of studies have shown miRNA mediated regulation of gene expression. To understand the complex network of miRNA:mRNA function, it is important to understand the molecular players of miRNA biogenesis and post-transcriptional processing. MicroRNA *Let-7* is the first known human miRNAs and is highly conserved in different organisms. Most of the human cancers show down-regulation of *Let-7* family members suggesting a tumor suppressive function of these miRNAs. *Let-7* has been shown to be an important candidate for predicting survival of cancer patients. For developing novel *Let-7* based therapeutic approaches, it is very important to understand the biogenesis and post-transcriptional regulation of this miRNA. This chapter summarizes the recent advances on regulatory molecules controlling transcriptional and post-transcriptional processing of *Let-7*.

2. INTRODUCTION

MiRNAs are small noncoding RNAs involved in post-transcriptional regulation of gene expression. MiRNAs are known to regulate gene expression either by degrading the mRNA or by inhibiting translation. They execute their gene regulatory function by associating with the RISC complex (Gregory et al., 2005) and translational machinery of the cell (Pillai et al.,

* Corresponding author: Neetu Dahiya, Department of Obstetrics and Gynecology, University of Virginia, Charlottesville, USA 22908.Email: ineetudahiya@yahoo.com.

2005). The mechanism of gene regulation i.e. mRNA degradation vs. translation inhibition by miRNA is determined by the complementarity of the target sequence where perfect complementarity favors degradation over translation inhibition. MiRNAs mediated fine tuning of gene expression and their involvement in regulation of numerous biological pathways makes them interesting candidates for studying their novel functions. Multiples studies have reported deregulation of miRNAs in various human diseases including cancer. Based on their expression pattern in cancer, miRNAs have been categorized as oncogenes or tumor suppressors. The immense potential of miRNAs as biomarker and therapeutic targets has encouraged researchers to study these important molecules. Understanding the regulation of miRNAs will provide further insight on the tightly regulated miRNA:mRNA network in the cell. This chapter summarizes the transcriptional and post-transcriptional regulation of *Let-7* in different organisms and important regulatory molecules involved in the process.

3. MiRNA Biogenesis

MiRNAs are transcribed as long primary transcript (pri-miRNA) by RNA polymerase II (Lee et al., 2004). The pri-miRNAs are further processed by Drosha-DGCR8 microprocessor complex into pre-miRNAs which are ~70bp in length. The pre-miRNAs are exported out of the nucleus by exportin 5 and cleaved by Dicer, a RNase III endonuclease enzyme into miRNA duplex in the cytoplasm. However, the recent discovery has suggested that pre-miRNAs can be processed by Dicer dependent or Dicer independent manner. Dicer independent pathway involves another member of RISC complex known as argonaute 2, which cleave the pre-miRNA passenger strand that undergo polyuridylation and nuclease mediated trimming to generate mature miRNA (Macfarlane et al., 2010). Once the mature miRNA is synthesized, one of the miRNA strands gets incorporated into RISC where it directs mRNA degradation of its targets. In addition to Dicer dependent and independent pathways, there is an alternative pathway called mirtron pathway which bypass Drosha processing. In mirtron pathway, pre-miRNA is generated by splicing and debranching of short hairpin introns called mirtrons. These small RNAs are exported to cytoplasm by exportin and further processed by Dicer. Most of the miRNA encoding sequences are present in the introns or exons of other genes and are transcribed together with the gene. However some miRNAs are intergenic and have independent promoters (McDermott et al., 2011). MiRNA expression seems to be regulated in a tissue and developmental specific manner (Zhao and Srivastava, 2007). The mechanism of miRNA regulation is further complexed by the fact that there are multiple steps in miRNA biogenesis which can be regulated by RNA polymerases such as Drosha, Dicer, ribonucleoprotein such as hnRNPA1 or Argonaute proteins such as Ago1, Ago2 or transcription factors such as SMAD1, SMAD5, LIN28, c-MYC, p53 (Heo et al., 2008; Hutvager et al., 2001; Zhang et al., 2009). In addition, miRNA expression is also regulated at epigenetic level (Rouhi et al., 2008).

4. *LET-7* MICRORNA

The *Let-7* was the second miRNA discovered in a genetic screen in C. elegans. There are eleven *Let-7* members identified so far (*Let-7*a-1, *Let-7*a-2, *Let-7*a-3, *Let-7*b, *Let-7*c, *Let-7*d, *Let-7*e, *Let-7*f-1, *Let-7*f-2, *Let-7*g, *Let-7*i) in miRBase Release 17. The *Let-7* family members are conversed among several species suggesting their critical role in essential cellular processes. In our previous study on miRNA profiling in ovarian cancer we found downregulation of several *Let-7* members (*Let-7*d, *Let-7*e and *Let-7*f) in ovarian cancer tissuses as well as cancer cell lines. Similar to ovarian cancer, downregulation of *Let-7* members has been reported in other cancers including gastric cancer (Ohshima et al., 2010), colorectal cancer (Ruzzo et al., 2011), glioblastoma (Lee et al., 2011), retinoblastoma (Mu et al., 2010), thyroid cancer (Ricarte-Filho et al., 2009), pancreatic cancer (Torrisani et al., 2009) and lung cancers (Takamizawa et al., 2004). Most of the *Let-7* members act as tumor suppressor by targeting oncogenes such as RAS, HMGA2 and c-MYC (Johnson et al., 2005; Lee and Dutta, 2007; Wong et al., 2011).

5. TRANSCRIPTIONAL REGULATION

MiRNA biogenesis can be regulated both at transcriptional or post-transcriptional level. Several molecules controlling transcription or post-transcriptional modification of *Let-7* have already been identified. Although there are several reports on post-transcriptional regulation of *Let-7* but not much information is available on its transcriptional regulation. Characterization of primary transcripts will help in understanding the structural elements and sequences involved in regulation of miRNA processing. Here are the different candidates regulating the transcription of *Let-7*.

1) DAF12

Let-7 shows temporal expression in animals including C. elegans. Johnson et al. (2003) identified TRE (translation repressing element) located ~1200bp upstream of the start site of mature miRNA. During early stages of development (L1-L3), the TRE is not activated so *Let-7* expression is low. However in L3/L4 transition, TRE is occupied by activators, thereby inducing expression of *Let-7*. The authors identified DAF12 as a positive regulator of *Let-7*, as mutations in DAF12 resulted in decreased expression of *Let-7* and retarded the heterochronic phenotype. The authors concluded that in C. elegans, *Let-7* is transcriptionally controlled by enhancer element which is activated by TREB factors such as DAF12 (Johnson et al., 2003). There is a feedback loop between *Let-7* family miRNAs and DAF12 which controls the developmental timings in *C. elegans* (Hammell et al., 2009).

2) CREBα

Guan et al. (2011) studied regulation of *Let-7*a in lung cancer cell line A549. They applied 5'RACE to characterize the promoter of human miRNA *Let-7*a-2. A 2.8 kb fragment from 5'flanking region of *Let-7*a-2 showed putative binding sites for transcription factors p53, C-MYC, RAS, CEBPα, PTEN, RORA, RXR, TCF and GR. Co-transfection of *Let-7*a-2 expressing vectors along with vectors expressing putative transcription identified CEBPα as a negative regulator of *Let-7*a-2 promoter activity.

3) HBL1

HBL1 is a zinc finger transcription factor involved in developmental regulation of C. elegans and a target of miRNA *Let-7*. *Let-7* promoter has a conserved 62bp region containing 3 putative HBL1 binding sites. These binding sites were in proximity of previously identified elements (TRE), starting ~18nt downstream of TRE element in *Let-7* promoter. HBL1 loss of function mutants showed increased expression of both pri-*Let-7* and mature *Let-7*, proving role of HBL1 in transcriptional regulation of *Let-7*.

6. Post-Transcription Regulation

The post-transcriptional regulation of *Let-7* has been widely studied. Here are the important candidate proteins and processes controlling post-transcriptional regulation of *Let-7*.

1) LIN28

LIN28, a RNA binding protein, is one of the main proteins involved in regulation *Let-7*. LIN28 contains -a cold-shock domain on N-terminus and two CCHC-type zinc finger domains at the C-terminus. LIN28 has known role in regulation of developmental timing in *C. elegans* similar to *Let-7*. LIN28 is involved in multilevel posttranscriptional regulation of *Let-7*. It interferes with nuclear and cytoplasmic *Let-7* processing. In addition, LIN28 inhibits *Let-7* processing at pri-*Let-7* stage. The expression of primary miRNA in mouse embryonic stem (ES) cells and embryonal carcinoma and human primary tumor but the lack of mature miRNA suggested a posttranscriptional regulation of miRNA. The pri-*Let-7*g was detectable in ES cells as well as after differentiation into embroid bodies. However, mature *Let-7*g was undetectable in undifferentiated ES cells but exhibited increased expression after differentiation (Viswanathan et al., 2008). To identify the factor responsible for blocking *Let-7* processing, the authors used a biochemical approach based on pre-*Let-7*g affinity binding. Mass spectrometry analysis identified several RNA binding proteins including LIN28. LIN28 was able to bind both pri- as well as pre-*Let-7*g. LIN28B, a homolog of LIN28, has been found upregulated in cancers. It exists in two different isoforms-LIN28B-S and LIN28B-L where the former carries a truncated cold shock domain. LIN28S is incapable of inhibiting processing of pri-*Let-7*g.

The precise mechanism of LIN28 mediated regulation of *Let-7* was revealed by studying effect of LIN28 modulation on *Let-7* expression combined with biochemical analysis. The siRNA mediated knockdown of LIN28B resulted in an increase in the mature *Let-7*a without affecting the pri-*Let-7*a suggesting a post-transcriptional regulation by LIN28 (Heo et al., 2008). Although the pre-*Let-7*a was also upregulated the effect was more pronounced on the mature *Let-7*a. All other tested *Let-7* members, *Let-7*g, *Let-7*b, *Let-7*d and *Let-7*e, exhibited the similar response to LIN28 knockdown. Since most of the LIN28 protein is localized in the cytoplasm and it did not affect the primary miRNA it was hypothesized that LIN28 might be regulating *Let-7* biogenesis after its export into the cytoplasm. In an immunoprecipitation based approach, LIN28 was found associated mainly with pre-*Let-7*f, less efficiently with mature *Let-7* and there was no association with pri-*Let-7* suggesting post-transcriptional regulation. In order to understand the mechanism of regulation, LIN28A, LIN28B and LIN28 mutant (with partially truncated cold-shock domain) were coexpressed with pri-*Let-7*a-1 in HEK293T cells which showed very low expression of endogenous LIN28 and *Let-7* (Heo et al., 2008). The level of mature *Let-7* was significantly low in presence of LIN28A and LIN28B cells however the cells expressing mutant LIN28 showed only a marginal change in the mature *Let-7*. LIN28 mutants (with point mutation in the zinc finger domains) were able to bind to pre-*Let-7* but did not affect the *Let-7* biogenesis suggesting a prominent role of ZFD in LIN28 mediated regulation of *Let-7*. Immunoprecipitation with LIN28A showed a band ~18nt longer than the pre-*Let-7*a which was not found in IP with LIN28 ZFD mutant confirming the functional role of ZFD domains. Sequencing and RNase H cleavage reaction confirmed the presence of U residues on larger pre-*Let-7*. The uridylated *Let-7* was resistant to dicer processing since dicer is unable to cleave the pre-miRNA with elongated 3' tail. Terminal uridylation of RNA serves as a signature for directing RNA decay (Heo et al., 2008). Since uridylated pre-*Let-7* degraded faster than the unmodified pre-*Let-7* it was proposed that LIN28 competes with dicer for pre-*Let-7*. It blocks *Let-7* biogenesis at dicer processing level, promotes pre-*Let-7* uridylation which ultimately leads to decay of pre-*Let-7*. LIN28 and *Let-7* regulates each other in a feedback loop system.

2) Dicer

Dicer is one of the main members of miRNA biogenesis pathway. Dicer contains an N-terminal DEXH-box ATP-dependent RNA helicase domain, PAZ domain, tandem RNase III motifs and C-terminal dsRNA-binding domain. siRNA mediated downregulation of Dicer resulted in accumulation of longer *Let-7* containing RNA similar to pre-*Let-7*. The mature *Let-7* was detectable in control siRNA treated cells but not in cells treated with Dicer siRNA providing evidence in supporting role of dicer in miRNA processing (Hutvágner et al., 2001). In another study, dicer protein expression was found inversely correlated with the expression of mature *Let-7*a in PC10 cells. To understand the mechanism of this regulation Dicer expression was analyzed after modulation of *Let-7* expression. Transfection of precursor *Let-7* in PC10 cells inhibited expression of Dicer at mRNA as well as at protein level. Further analysis identified, Dicer as a direct target of *Let-7* harboring *Let-7* binding site in the 3'UTR. In addition to 3'UTR, the *Let-7* binding sites in the coding region of Dicer are also functional and involved in *Let-7* mediated downregulation of Dicer protein (Tokumaru et al., 2008). This is a proof of miRNA/Dicer autoregulatory negative feedback loop. However, it was suggested that the *Let-7* binding sites in 3'UTR and coding region may differ in the mechanism of regulation. The mechanism was dissected by comparing base-pairing probabilities along the entire *Let-7* miRNA with conserved and nonconserved target seeds 3'UTR and coding region. In 3'UTR, conserved sites showed preference for looping at base pairs 10-12 and binding at positions 13-16 compared to non-conserved sites. On the other hand, conserved binding sites in coding region preferred binding at positions 13-15 and 17-19 but no preference for looping, suggesting a possible mechanistic difference between binding sites in 3'UTR and coding region.

Lightfoot et al., (2011) identified a region in the terminal loop of precursor *Let-7*g that interacts with LIN28 directly. Binding of LIN28 induces conformational changes in pre-*Let-7*g loop resulting in unwinding of the upper stem region of pre-*Let-7*g at U21/U22. Dicer cleavage site is located at U22 and a double stranded structure is thought to be essential for *Let-7* maturation by Dicer (Hutvágner et al., 2001). This proves direct role of LIN28 in *Let-7* processing at Dicer step. To prove this concept, Hutvagner et al. (2001) used a mutant pre-*Let-7*g (where G20, U21 and U22 were mutated to A) which has open upper stem. In an *in-vitro* Dicer processing assay, wild type pre-*Let-7* was processed to mature *Let-7*g but in mutated pre-*Let-7*g the amount of mature *Let-7*g was very low. The addition of LIN28 inhibited dicer mediated processing of wild type pre-*Let-7*g. But in the absence of Dicer, LIN28 had no effect on wild type pre-*Let-7*g.

NF-kB is involved in indirect regulation of *Let-7*. It binds to first intron of LIN28 and activates it. Increased expression of LIN28 inhibits *Let-7* expression through posttranscriptional mechanism (Iliopoulos et al., 2009). Recently, miR-181 is identified as a novel regulator of LIN28-*Let-7* feedback loop by directly targeting LIN28 levels and subsequent effects on *Let-7* levels during megakaryocytic differentiation (Li et al., 2011).

3) Drosha

Newman et al., (2008) dissected the mechanism of LIN28 mediated repression of *Let-7*. Using Drosha inhibitor, the authors identified conserved nucleotides in the loop region of primary transcripts that are essential for the blocking Drosha processing. LIN28 binds to the *Let-7* loop region and this binding is specific for *Let-7* members. LIN28 binding blocks the binding of Drosha and interfere with processing of primary *Let-7* transcript to pre-*Let-7*. Although LIN28 binds to *Let-7* transcript, the exact mechanism of *Let-7* regulation is unknown. The various reasons may be i) block access to Drosha, degradation of pri-miRNA if not processed by Drosha, ii) export of LIN28/*Let-7* complex to P-bodies for degradation, the latter is supported by localization of LIN28 in nucleus as well as in P-bodies (Newman et al., 2008).

LIN28 binds the loop region of *Let-7* miRNA. The further details of the mechanism were studied by Piskounova et al. (2008). The molecular determinants of LIN28 mediated miRNA processing were determined using EMSA. LIN28 binds to the terminal loop of the pre-*Let-7* miRNA (this region is present in both the pri-*Let-7* and pre-*Let-7*) and the binding site harbors a conserved cytosine nucleotide whose alteration affects LIN28 binding even without affecting the folding of the pre-*Let-7*g loop region (Piskounova et al 2008).

4) Argonaute

Argonaute proteins are key effectors of small RNA-mediated regulatory pathways. Argonaute 2 (also known as Ago2) proteins are well conserved and contain an amino-terminal domain, mid-domain, PAZ domain and Piwi domains. Among different Ago members, Ago2 exhibits endonuclease activity (Cheloufi et al., 2010). Stable overexpression of Ago2 in 293T cells showed modest increase in Let-7 miRNA. The authors suggested a feedback loop involving Let-7, LIN28 and c-MYC. Ago2 mediated upregulation of *Let-7* inhibited expression of LIN28 and c-MYC. Although the exact mechanism of regulation is not known, it was suggested to inhibit expression of LIN28 which interfere with *Let-7* processing at Dicer and Drosha level (Zhang et al., 2009).

5) FAS

FAS, a cell surface receptor is a member of tumor necrosis factor receptor family showing higher expression in normal cells compared to cancers. Interferon-gamma is involved in FAS-mediated cell death in cancers such as ovarian, renal and prostate cancers (Kim et al., 2002; Tomita et al., 2003; Selleck et al., 2003). To identify the role of miRNAs in FAS mediated cell death, the authors looked for predicted miRNAs targeting FAS and found *Let-7*/miR98 family members as potential regulators of FAS. They found an inverse correlation between FAS and *Let-7* expression in several cancer cell lines. FAS was identified as direct target of *Let-7* in dual-luciferase reporter assay. Co-transfection of luciferase vectors carrying FAS 3'UTR with *Let-7* binding sites showed reduced expression of luciferase compared to the plasmid carrying FAS 3'UTR with mutated *Let-7* binding sites. INF-gamma treatment induced FAS expression in HT29 cells and decreased expression of *Let-7* family members again confirming the negative correlation between FAS and *Let-7*. Further analysis showed reduced expression of both pri-*Let-7* and pre-*Let-7* after IFN-gamma treatment indicating FAS mediated regulation of *Let-7* at both transcriptional and post-transcriptional level (Geng et al., 2011).

6) RNA Editing/Splicing

RNA editing is a process involving nucleoside modifications, addition or deletion of which alters the aminoacid sequence of the encoded protein. RNA editing generates RNA and protein diversity by altering coding and non coding sequences (Luciano et al., 2004). RNA editing has been already reported in eukaryotic tRNAs, rRNAs, mRNAs and microRNA. The RNA editing enzyme, known as the adenosine deaminases acting on RNA (ADARs) specifically recognize partially double stranded RNA structures and modify the adenosine residues. The A-to-I RNA editing has been reported to affect the regulation of pri- and pre-miRNA processing (Bass, 2002). In addition to terminal well known U-and A-insertions, several internal mismatches have been reported in the some miRNAs. However, the Next generation sequencing has revolutionized the field of miRNAs helping in understanding the novel aspects of miRNA biogenesis, processing and functions. With parallel sequencing it is now possible to find out the internal mismatches in the RNAs in larger number of samples.

The functional significance of RNA editing in terms of miRNAs mediated target degradation is not well known. But it may affect the target degradation pathway by altering the complementarity between mRNA:miRNA complex. Since mRNAs exhibiting perfect complementarity to miRNA seed region are degraded more efficiently and mRNA with less complementarity are regulated at translation level, RNA editing may shift the mRNA regulation pathway from the target cleavage depending on the complementarity between the edited miRNA and the target mRNA (Reid et al., 2008). In addition, RNA editing in non-coding RNAs can lead to alternative splicing and stability of the RNAs.

1. Regulation by Polyuridylation Enzymes

Polyuridylation of *Let-7* miRNA is suggested as another mode of posttranscriptional regulation of *Let-7* miRNA. The presence of uridine residues recruits exonucleases (Mullen and Marzluff, 2008) which cleave the RNA molecule. Two polyuridylating enzymes have been reported to add uridine residues to *Let-7* transcripts.

a) TUT4

During miRNA biogenesis, primary miRNA transcript is processed by Drosha into a hairpin RNA called pre-miRNA. The pre-miRNA is processed to mature miRNA by Dicer. However presence of several uridine residues at 3'end of the pre-miRNAs interferes with Dicer processing, subsequently resulting in a decreased level of mature miRNA. LIN28 doesn't have uridylation activity. It may recruit some uridylating enzyme. RNA affinity purification by *Let-7*a in 293T cells overexpressing LIN28 followed by Mass spectrometry analysis identified TUT4 as another binding protein. It is also confirmed in another cell line Huh7, siRNA mediated downregulation of TUT4 led to increased expression of mature *Let-7* miRNAs. However, TUT4 associates with *Let-7* only in presence of LIN28. After siRNA mediated downregulation of LIN28, TUT4 failed to associate with *Let-7*. TUTase4, terminal uridylyl transferase 4 or ZCCHC11 was recently shown to be involved in the regulation of miRNA by adding uridine residues to the pre-miRNA (Heo et al., 2009). However Tut4 was unable to bind to *Let-7* or LIN28 in absence of the other partner suggesting requirement of both LIN28 and pre-*Let-7* for binding of Tut4. In colon cancers *Let-7* miRNAs are decreased compared to the adjacent normal cells. However, pri-*Let-7* levels are unchanged suggesting a post-transcriptional regulation of *Let-7* miRNA. The *Let-7* and LIN28 might be involved in a feedback loop. c-MYC in increased in colon cancer cells which may transactivate LIN28B which regulates *Let-7*. On the other hand LIN28 is a target of miRNA *Let-7*, this feedback loop is tightly regulated (King et al., 2011). MYC overexpression affects miRNA biogenesis, mostly by downregulation of miRNAs at transcription level suggested by low level of primary miRNA transcript. Although *Let-7* primary transcripts were unaffected, mature *Let-7*g was down two fold in MYC expressing cells. MYC induces LIN28 and chromatin IP and reporter assay shows direct association of MYC with LIN28 promoter resulting in transcriptional transactivation. siRNA mediated LIN28 downregulation completely reversed MYC mediated *Let-7* repression of mature *Let-7*g (Pan et al., 2011)

b) PUP2

Analysis of high-throughput sequencing data indicated in vivo uridylation of *Let-7*. LIN28 mediates uridylation of pre-*Let-7* followed by degradation in mammalian cell lines

(Heo et al., 2008). Since LIN28 cannot add uridine residues to the RNA moleculae, Lehrbach et al. (2009) applied RNAi based screening against 15 potential uridylating enzyme and checked their effect on level of mature and pre-*Let-7*. Downregulation of PUP2 resulted in an increase in pre-*Let-7* and a small increase in mature *Let-7*, suggesting its role in LIN28 mediated blockade of *Let-7* processing. In C. elegans, LIN28 recruits poly (U) polymerase PUP2 to uridylate pre-*Let-7* (Lehrbach et al., 2009).

2) Regulation by Splicing Factors

a) HnRNP A1 and KSRP

During miRNA biogenesis, miRNAs are processed in microprocessor complex which consists of RNase III enzyme Drosha, a double stranded RNA binding protein DGCR8, Dead Box helicases such as p68/72 and several hnRNP proteins. The microprcessor complex carrying only Drosha and DGCR8 is sufficient for miRNA processing (Michlewski and Cáceres, 2010) therefore, the other members of the microprocessor complex were suggested to have other regulatory roles. HnRNP A1, a nucleocytoplasmic shuttling protein is known to have role in miRNA metabolism by introducing conformational changes. It binds to conserved terminal loop of the *Let-7*a precursor and blocks its Drosha mediated processing in somatic cells where LIN28 is not expressed. Another splicing protein, KSRP (known as KH-type splicing regulatory protein) is a component of Drosha and Dicer complexes involved in regulatory roles in miRNA biogenesis. Terminal loop of *Let-7*a-1 has two GGG triplets which are optimal binding sites for KSRP. In vitro processing of pri-*Let-7*a having mutation in GGG in its terminal loop was significantly reduced compared to wild type suggesting positive effect of KSRP on *Let-7*a processing. The introduction of GGG mutation also abrogates binding of HnRNP A1 but the lack of binding of KSRP is dominant.

LIN28 negatively regulates *Let-7* biogenesis whereas KSRP promotes *Let-7* biogenesis and LIN28 dominates over KSRP. However, cells lacking LIN28 and having similar expression of KSRP showed difference in the levels of mature *Let-7*, suggesting alternative regulatory mechanism(s). Since HnRNP A1 can bind to the terminal loop of *Let-7* and there was a negative correlation between expression of HnRNPA1 and *Let-7* level in 293T cells, HnRNPA1 was suggested a regulator of *Let-7*. This regulation was further confirmed by ectopic expression of HnRNPA1, which resulted in reduced expression of *Let-7*a in two different cell lines. In order to identify the inhibitory effect of HnRNPA1 on *Let-7* biogenesis, the authors assessed the expression of pri-*Let-7*a-1 and mature *Let-7*a. Increased expression of HnRNPA1 was associated with accumulation of primary *Let-7*a transcript, suggesting a posttranscriptional regulation of *Let-7* in Drosha mediated processing. In the *in-vitro* assay, HnRNPA1 inhibited pri-*Let-7*a processing in a dose dependent manner. On the other hand, KSRP showed a stimulatory effect on pri-*Let-7*a-1 processing suggesting antagonistic roles of hnRNP A1 and KSRP in biogenesis of *Let-7*.

3) Transplicing/Polyadenylation

MiRNAs genes are present either as independent transcriptional unit or within annotated host genes. In addition more than 50% miRNAs genes are present in the introns (Morlando et al., 2008). They are either transcribed from their independent promoter or by the promoter of the host gene. To understand the miRNA biogenesis and identify important elements controlling miRNA gene expression, there is need to characterize the complete primary transcripts. Bracht et al. (2004) focused on *Let-7* miRNAs in C. elegans and proposed two primary transcripts ~1731-nt and ~890-nt which are processed to mature *Let-7*. These transcripts exhibited features of Pol-II directed synthesis showing presence of 5'-capped and 3'-end polyadenylated and substrates for splicing. The sequences flanking the miRNA hairpin may regulate the pri-miRNA processing to mature miRNA. Although, the exact function of splicing in pri-*Let-7* miRNA processing is not known, it may remove some inhibitory sequences or change structural elements surrounding the hairpin precursor (Bracht et al., 2004).

Conclusion

Let-7 is one of the most important miRNAs involved in development and cancer. *Let-7* expression is tightly regulated at transcriptional and post-transcriptional level. DAF12 and HBL1 regulate *Let-7* expression by inhibiting transcription and exists in a feedback loop. The post-transcriptional regulation of *Let-7* has been widely studied. The most important post-transcriptional regulator of *Let-7* is LIN28, showing regulation at both pri-miRNA and pre-miRNA processing. In addition to interfering with Dicer and Drosha level, LIN28 plays important role in TUT4 and PUP2 mediated uridylation of *Let-7* and the subsequent reduction in *Let-7* expression. Other molecules such as FAS, c-MYC, HnRNP A1 are also negative regulators of *Let-7* expression whereas KSRP and Ago2 are positive regulator of *Let-7* processing.

References

Bass, B. L. RNA editing by adenosine deaminases that act on RNA. *Annu. Rev. Biochem.* 2002;71:817-846.

Bracht, J., Hunter, S., Eachus, R., Weeks, P., Pasquinelli, A. E. Trans-splicing and polyadenylation of *Let-7* microRNA primary transcripts. *RNA*. 2004;10:1586-1594.

Cheloufi, S., Dos Santos, C. O., Chong, M. M., Hannon, G. J. A dicer-independent miRNA biogenesis pathway that requires Ago catalysis. *Nature*. 2010;465:584-589.

Geng, L., Zhu, B., Dai, B. H., Sui, C. J., Xu, F., Kan, T., Shen, W. F., Yang, J. M. A *Let-7*/FAS double-negative feedback loop regulates human colon carcinoma cells sensitivity to FAS-related apoptosis. *Biochem. Biophys. Res. Commun.* 2011;408:494-499.

Gregory, R. I., Chendrimada, T. P., Cooch, N., Shiekhattar, R. Human RISC couples microRNA biogenesis and posttranscriptional gene silencing. *Cell* 2005;123:631-640.

Guan, H., Zhang, P., Liu, C., Zhang, J., Huang, Z., Chen, W., Chen, Z., Ni, N., Liu, Q., Jiang, A. Characterization and functional analysis of the human microRNA *Let-7*a2 promoter in lung cancer A549 cell lines. *Mol. Biol. Rep.* 2011; 38:5327-5334.

Hammell, C. M., Karp, X., Ambros, V. A feedback circuit involving *Let-7*-family miRNAs and DAF-12 integrates environmental signals and developmental timing in Caenorhabditis elegans. *Proc. Natl. Acad. Sci.* U S A. 2009;106:18668-18673.

Heo, I., Joo, C., Cho, J., Ha, M., Han, J., Kim, V. N. LIN28 mediates the terminal uridylation of *Let-7* precursor MicroRNA. *Mol. Cell.* 2008;32:276-284.

Heo, I., Joo, C., Kim, Y. K., Ha, M., Yoon, M. J., Cho, J., Yeom, K. H., Han, J., Kim, V. N. TUT4 in concert with LIN28 suppresses microRNA biogenesis through pre-microRNA uridylation. *Cell.* 2009;138:696-708.

Hutvágner, G., McLachlan, J., Pasquinelli, A. E., Bálint, E., Tuschl, T., Zamore, P. D. A cellular function for the RNA-interference enzyme Dicer in the maturation of the *Let-7* small temporal RNA. *Science.* 2001;293:834-838.

Iliopoulos, D., Hirsch, H. A., Struhl, K. An epigenetic switch involving NF-kappaB, LIN28, *Let-7* MicroRNA, and IL6 links inflammation to cell transformation. *Cell.* 2009;139:693-706.

Johnson, S. M., Grosshans, H., Shingara, J., Byrom, M., Jarvis, R., Cheng, A., Labourier, E., Reinert, K. L., Brown, D., Slack, F. J. RAS is regulated by the *Let-7* microRNA family. *Cell.* 2005;120:635-647.

Johnson, S. M., Lin, S. Y., Slack, F. J. The time of appearance of the C. elegans *Let-7* microRNA is transcriptionally controlled utilizing a temporal regulatory element in its promoter. *Dev. Biol.* 2003;259:364-379.

Kim, E. J., Lee, J. M., Namkoong, S. E., Um, S. J., and Park, J. S. Interferon regulatory factor-1 mediates interferon-gamma-induced apoptosis in ovarian carcinoma cells. *J. Cell Biochem.* 2002;85:369-380.

King, C. E., Wang, L., Winograd, R., Madison, B. B., Mongroo, P. S,. Johnstone, C. N., Rustgi, A. K. LIN28B fosters colon cancer migration, invasion and transformation through *Let-7*-dependent and -independent mechanisms. *Oncogene.* 2011;30:4185-4193.

Lee, S. T., Chu, K., Oh, H. J., Im, W. S., Lim, J. Y., Kim, S. K., Park, C. K., Jung, K. H., Lee, S. K., Kim, M., Roh, J. K. *Let-7* microRNA inhibits the proliferation of human glioblastoma cells. *J. Neurooncol.* 2011;102:19-24.

Lee, Y., Kim, M., Han, J., Yeom, K. H,. Lee, S., Baek, S. H., Kim, V. N. MicroRNA genes are transcribed by RNA polymerase II. *EMBO J.* 2004;23:4051-4060.

Lee, Y. S., Dutta, A. The tumor suppressor microRNA *Let-7* represses the HMGA2 oncogene. *Genes Dev.* 2007;21:1025-1030.

Lehrbach, N. J., Armisen, J., Lightfoot, H. L., Murfitt, K. J., Bugaut, A., Balasubramanian, S., Miska, E. A. LIN28 and the poly(U) polymerase PUP-2 regulate *Let-7* microRNA processing in Caenorhabditis elegans. *Nat. Struct. Mol. Biol.* 2009;16:1016-1020.

Li, X., Zhang, J., Gao, L., McClellan, S., Finan, M. A., Butler, T. W., Owen, L. B., Piazza, G. A., Xi, Y. MiR-181 mediates cell differentiation by interrupting the LIN28 and *Let-7* feedback circuit. *Cell Death Differ.* 2011.

Lightfoot, H. L., Bugaut, A., Armisen, J., Lehrbach, N. J., Miska, E. A., Balasubramanian, S. A LIN28-dependent structural change in pre-*Let-7*g directly inhibits dicer processing. *Biochemistry.* 2011;50:7514-7521.

Luciano, D. J., MiRsky, H., Vendetti, N. J., Maas, S. RNA editing of a miRNA precursor. *RNA.* 2004;10:1174-1177.

Macfarlane, L. A., Murphy, P. R. MicroRNA: Biogenesis, Function and Role in Cancer. *Curr. Genomics.* 2010;11:537-561.

McDermott, A. M., Heneghan, H. M., Miller, N., Kerin, M. J. The Therapeutic Potential of MicroRNAs: Disease Modulators and Drug Targets. *Pharm. Res.* 2011

Michlewski, G., Cáceres, J. F. Antagonistic role of hnRNP A1 and KSRP in the regulation of *Let-7*a biogenesis. *Nat. Struct. Mol. Biol.* 2010;17:1011-1018.

Morlando, M., Ballarino, M., Gromak, N., Pagano, F., Bozzoni, I., Proudfoot, N. J. Primary microRNA transcripts are processed co-transcriptionally. *Nat. Struct. Mol. Biol.* 2008;15:902-909.

Mu, G., Liu, H., Zhou, F., Xu, X., Jiang, H., Wang, Y., Qu, Y. Correlation of overexpression of HMGA1 and HMGA2 with poor tumor differentiation, invasion, and proliferation associated with *Let-7* down-regulation in retinoblastomas. *Hum. Pathol.* 2010;41:493-502.

Mullen, T. E., Marzluff, W. F. Degradation of histone mRNA requires oligouridylation followed by decapping and simultaneous degradation of the mRNA both 5' to 3' and 3' to 5'. *Genes Dev*. 2008;22:50-65.

Newman, M. A., Thomson, J. M., Hammond, S. M. LIN28 interaction with the *Let-7* precursor loop mediates regulated microRNA processing. *RNA*. 2008;14:1539-1549.

Ohshima, K., Inoue, K., Fujiwara, A., Hatakeyama, K., Kanto, K., Watanabe, Y., Muramatsu, K., Fukuda, Y., Ogura, S., Yamaguchi, K., Mochizuki, T. *Let-7* microRNA family is selectively secreted into the extracellular environment via exosomes in a metastatic gastric cancer cell line. *PLoS One*. 2010;5:e13247.

Pan, L., Gong, Z., Zhong, Z., Dong, Z., Liu, Q., Le, Y., Guo, J. LIN28 reactivation is required for *Let-7* repression and proliferation in human small cell lung cancer cells. *Mol. Cell Biochem*. 2011;355:257-263.

Pillai, R. S., Bhattacharyya, S. N., Artus, C. G., Zoller, T., Cougot, N., Basyuk, E., Bertrand, E., Filipowicz, W. Inhibition of translational initiation by *Let-7* MicroRNA in human cells. *Science*. 2005;309:1573-1576.

Piskounova, E., Viswanathan, S. R., Janas, M., LaPierre, R. J., Daley, G. Q., Sliz, P., Gregory, R. I. Determinants of microRNA processing inhibition by the developmentally regulated RNA-binding protein LIN28. *J. Biol. Chem*. 2008;283:21310-21314.

Reid, J. G., Nagaraja, A. K., Lynn, F. C., Drabek, R. B., Muzny, D. M., Shaw, C. A., Weiss, M. K., Naghavi, A. O., Khan, M., Zhu, H., Tennakoon, J., Gunaratne, G. H., Corry, D. B., Miller, J., McManus, M. T., German, M. S., Gibbs, R. A., Matzuk, M. M., Gunaratne, P. H. Mouse *Let-7* miRNA populations exhibit RNA editing that is constrained in the 5'-seed/ cleavage/anchor regions and stabilize predicted mmu-*Let-7*a:mRNA duplexes. *Genome Res*. 2008;18:1571-1581.

Ricarte-Filho, J. C., Fuziwara, C. S., Yamashita, A. S., Rezende, E., Da-Silva, M. J., Kimura, E. T. Effects of *Let-7* microRNA on Cell Growth and Differentiation of Papillary Thyroid Cancer. *Transl. Oncol*. 2009;2:236-241.

Rouhi, A., Mager, D. L., Humphries, R. K., Kuchenbauer, F. MiRNAs, epigenetics, and cancer. *Mamm. Genome*. 2008;19:517-525.

Roush, S. F., Slack, F. J. Transcription of the C. elegans *Let-7* microRNA is temporally regulated by one of its targets, hbl-1. *Dev. Biol.* 2009;334:523-534.

Ruzzo, A., Canestrari, E., Galluccio, N., Santini, D., Vincenzi, B., Tonini, G., Magnani, M., Graziano, F. Role of KRAS *Let-7* LCS6 SNP in metastatic colorectal cancer patients. *Ann. Oncol*. 2011;22:234-5.

Selleck, W. A., Canfield, S. E., Hassen, W. A., Meseck, M., Kuzmin, A. I., Eisensmith, R. C., Chen, S. H., and Hall, S. J. IFN-gamma sensitization of prostate cancer cells to FAS-mediated death: a gene therapy approach. *Mol. Ther*. 2003;7:185-192.

Takamizawa, J., Konishi, H., Yanagisawa, K., Tomida, S., Osada, H., Endoh, H., Harano, T., Yatabe, Y., Nagino, M., Nimura, Y., Mitsudomi, T., Takahashi, T. Reduced expression of the *Let-7* microRNAs in human lung cancers in association with shortened postoperative survival. *Cancer Res*. 2004;64:3753-3756.

Tokumaru, S., Suzuki, M., Yamada, H., Nagino, M., Takahashi, T. *Let-7* regulates Dicer expression and constitutes a negative feedback loop. *Carcinogenesis*. 2008; 29:2073-2077.

Tomita, Y., Bilim, V., Hara, N., Kasahara, T., and Takahashi, K. Role of IRF-1 and caspase-7 in IFN-gamma enhancement of FAS-mediated apoptosis in ACHN renal cell carcinoma cells. *Int. J. Cancer*. 2003;104: 400-408.

Torrisani, J., Bournet, B., Du Rieu, M. C., Bouisson, M., Souque, A., Escourrou, J., Buscail, L., Cordelier, P. *Let-7* MicroRNA transfer in pancreatic cancer-derived cells inhibits in vitro cell proliferation but fails to alter tumor progression. *Hum. Gene Ther*. 2009;20:831-844.

Viswanathan, S. R., Daley, G. Q. and Gregory, R. I. Selective blockade of microRNA processing by LIN28. *Science*. 2008;320:97-100.

Wong, T. S., Man, O. Y., Tsang, C. M., Tsao, S. W., Tsang, R. K., Chan, J. Y., Ho, W. K., Wei, W. I., To, V. S. MicroRNA *Let-7* suppresses nasopharyngeal carcinoma cells proliferation through downregulating c-MYC expression. *J. Cancer Res. Clin. Oncol*. 2011;137:415-422.

Zhang, X., Graves, P. R., Zeng, Y. Stable Argonaute2 overexpression differentially regulates microRNA production. *Biochim. Biophys. Acta*. 2009;1789:153-159.

Zhang, X., Graves, P. R., Zeng, Y. Stable Argonaute2 overexpression differentially regulates microRNA production. *Biochim. Biophys. Acta*. 2009;1789:153-159.

Zhao, Y., Srivastava, D. A developmental view of microRNA function. *Trends Biochem. Sci*. 2007;32:189-197.

In: MicroRNA *Let-7*
Editor: Neetu Dahiya

ISBN: 978-1-62081-152-8

Chapter 3

LET-7 AND SIGNALING PATHWAYS IN CANCERS

Yuanjia Tang[1]* and Nan Shen
[1]Joint Molecular Rheumatology Laboratory of Institute of Health Sciences and Shanghai Renji Hospital, Shanghai JiaoTong University School of Medicine and Shanghai Institutes for Biological Sciences, Chinese Academy of Sciences, Shanghai, China

1. ABSTRACT

The *Let-7* family is one of the first identified miRNA families, and is highly conserved across animal species in sequence. The *Let-7* family members have identical seed sequences and may overlap sets of targets. *Let-7* interwines with cellular networks to exert its function. *Let-7* has been identified as cancer repressor or oncogene. In this chapter, we focus on discussion of *Let-7* and cancer signaling. Particular *Let-7* family members have been shown to be specifically deregulated in certain cancers. Epigenetic regulation, and processing at transcription and post-transcriptional level are probably involved in *Let-7* deregulation in cancers. Deregulation of *Let-7* may contribute to pathological process by changing the threshold of apoptosis and cell cycle checkpoints in a tissue or stage dependent manner. The *Let-7* family regulates not only a single important pathway protein, but rather coordinate gene expression levels on a pathway-wide scale. *Let-7*-regulated genes were found enriched in focal adhesion, apoptosis, MAPK and cell cycle signaling pathways. Manipulation of *Let-7* expression may provide a potential strategy for designing novel anti-cancer therapies.

2. INTRODUCTION

The precise level of genes expression and normally controlled signaling pathway are essential for normal cell specification and tissue growth. The same core signaling components

* Corresponding author: Joint Molecular Rheumatology Laboratory of Institute of Health Sciences, Chinese Academy of Sciences, Shanghai, China. E-mail: yjtang@sibs.ac.cn.

are configured to generate very different transcriptional outputs in specific temporal or spatial settings. As a consequence, dysfunction of these components invariably induces oncogenesis. The fundamental control of cell division, growth and apoptosis by core signaling pathways makes many of their components potent oncogenes and tumor suppressors (Hagen and Lai, 2008). MicroRNAs (miRNAs) have emerged as central posttranscriptional regulators of gene expression, which can exert the turning-off function at post transcriptional level through repressing mRNA translation and/or mediating cleavage of mRNAs. MiRNA targets include signaling proteins, metabolic enzymes, transcription factors and so on, which appears to form a complex regulatory network. MiRNAs can be regarded as regulators that regulate all kinds of cellular networks and can also be regarded as new layers of signaling regulator. MiRNAs have a high spatio-temporal expression, suggesting that miRNAs may play an important role in the regulation of the strength and specificity of cellular signaling pathways through controlling the concentration of signal transduction components at different temporal and spatial conditions. It is reasonable to think that deregulation of miRNAs results in abnormal signaling pathways, which contributes to human diseases such as cancers. MiRNAs function as oncogenes or tumor suppressors during tumor development and progression.

The *Let-7* family is one of the first identified miRNA families, which consists of 11 *Let-7* members (*Let-7*a-1, *Let-7*a-2, *Let-7*a-3, *Let-7*b, *Let-7*c, *Let-7*d, *Let-7*e, *Let-7*f-1, *Let-7*f-2, *Let-7*g, *Let-7*i) in human beings and is highly conserved across animal species in sequence. The *Let-7* family members have identical seed sequences and may overlap sets of targets. *Let-7* has been identified as cancer repressor or oncogene. In this chapter, we focus on discussion of *Let-7* and its role in cancer signaling.

3. *LET-7* DEREGULATION IN CANCER

Loss- or gain-of-function of specific miRNAs appears to be a key event in the genesis of many human cancers. MiRNA expression profiling studies showed that miRNA expression levels are altered in human cancers, suggesting that miRNA profiling has diagnostic and perhaps prognostic potential. Deregulation of *Let-7* has been reported in a number of human cancers. Human cancer-associated *Let-7* family members were obtained from the PhenomiR database (http://mips.helmholtz-muenchen.de/phenomir/). The PhenomiR database provides information about differentially regulated miRNA expression in diseases and other biological processes (Ruepp et al., 2010). We used patient study data obtained from cancer related research and found the *Let-7* family members are found to be downregulated (Table 1) or upregulated (Table 2) in various human cancers including lung, colon, ovarian, breast cancer, leukemia and prostate cancer.

Let-7 is widely viewed as a tumor suppressor. The loss of *Let-7* family members also has prognostic value as it indicates poor survival in lung cancer and ovarian cancer (Takamizawa et al., 2004; Shell et al., 2007). Upregulation of certain *Let-7* family members has also been observed in several cancers. The upregulation of *Let-7*b and *Let-7*i was associated with high grade transformation in lymphoma (Lawrie et al., 2009). Hypomethylation of the *Let-7*a-3 locus was found to cause higher expression of *Let-7*a-3 in lung cancer (Brueckner et al., 2007). The conflicting data on the deregulation of *Let-7* in various cancers suggests that *Let-7* may have different functions in different tissues or in different cell contexts.

4. *LET-7* FUNCTIONS AS A NETWORK COMPONENT IN CANCER SIGNALING PATHWAY

Altered *Let-7* levels might result in inaccurate target protein levels in cell or tissue, consequently immoderate signal transduction, and contribute to the pathogenesis of cancer. Growing evidence demonstrates that restoration of *Let-7* expression can normalize the gene regulatory network and signaling pathways, and reverse the phenotype in cancerous cells. The *Let-7* family regulates not only a single important pathway protein, but rather coordinate gene expression levels on a pathway-wide scale. We got hundreds of *Let-7*-regulated genes (LRGs) via the miRWalk database, a comprehensive database that provides information on miRNA from Human, Mouse and Rat on their predicted as well as validated binding sites on their target genes (Dweep et al., 2011). Since *Let-7* has hundreds of targets, it may have complex influences on multiple components in one pathway, and perhaps even on components of different signaling pathways. LRGs were analyzed with the composite regulatory signature database (CRSD) (Liu et al., 2006) and were found enriched in several cancer related signaling pathway, such as focal adhesion, apoptosis, MAPK and cell cycle. In the following, we discussed the signaling pathway regulated by *Let-7*.

Table 1. Downregulated *Let-7* family members expression in human cancers

Cancer types	*Let-7* family members	Reference
Breast cancer	*Let-*7a,7d,7f	Iorio et al., 2005
	*Let-*7a	Sempere et al., 2007
	*Let-*7a	Volinia et al., 2006
	*Let-*7a,7c,7g	Zhang et al., 2006
Colorectal cancer	*Let-*7a,7b,7c,7f,7g	Cummins et al.,2006
	*Let-*7a	Fang et al., 2007
	*Let-*7a	Akao et al.,2006
	*Let-*7e	Monzo et al., 2008
Gastric cancer	*Let-*7a	Zhang et al.,2007
Hepatocellular carcinoma	*Let-*7a,7b,7c,7d,7e,7f,7g	Gramantieri et al.,2007
	*Let-*7g	Budhu et al., 2008
	*Let-*7c	Pineau et al., 2010
Leukemia, acute myeloid	*Let-*7b,7c,7e	Jongen et al.,2008
	*Let-*7a,7g	Cammarata et al., 2010
Lung cancer	*Let-*7a,7f	Takamizawa et al., 2004
	*Let-*7a,7c	Johnson et al., 2005
	*Let-*7a	Yanaihara et al., 2006
	*Let-*7a	Peltier et al., 2008
	miR-98,*Let-*7a,7f,7g,7i	Chen et al., 2008
Ovarian cancer	*Let-*7a,7b	Zhang et al., 2006
	miR-98,*Let-*7a,7c,7d,7f	Iorio et al., 2007
	*Let-*7a,7b,7c	Yang et al., 2008
	*Let-*7b	Nam et al., 2008
	*Let-*7e,7f	Dahiya et al., 2008
Pancreatic cancer	*Let-*7a	du Rieu et al., 2010
Prostate cancer	*Let-*7a	Volinia et al., 2006
	*Let-*7a,7b,7c,7d,7f,7g	Porkka et al., 2007

Table 2. Upregulated *Let-7* family members expression in human cancers

Cancer types	*Let-7* family members	Reference
Breast cancer	*Let-7*i	Iorio et al., 2005
	*Let-7*a,7b,7e	Zhang et al., 2006
Colorectal cancer	*Let-7*g	Nakajima et al., 2006
	miR-98,*Let-7*d,7g,7i	Monzo et al., 2008
Hepatocellular carcinoma	miR-98,*Let-*7a,7b,7c,7d,7e,7f,7g,7i	Huang et al., 2008
	*Let-7*a,7b,7d,7f,7i	Jiang et al., 2008
	*Let-7*g	Varnholt et al., 2008
Leukemia, acute myeloid	miR-98,*Let-*7a,7c,7d,7f,7g	Garzon et al., 2008
	*Let-7*b,7e	Jongen et al., 2008
	*Let-7*e	Dixon et al., 2008
	*Let-7*b	Cammarata et al.,2010
Lung cancer	*Let-7*a,7g	Volinia et al., 2006
	*Let-7*b,7c,7d,7e	Chen et al., 2008
Neuroblastoma	*Let-7*a,7b	Chen et al., 2007
	*Let-7*b	Schulte et al., 2008
Ovarian cancer	*Let-7*g,7i	Laios et al., 2008
Pancreatic cancer	*Let-7*d,7f	Lee et al., 2007
Prostate cancer	*Let-7*d,7i	Volinia et al., 2006
	*Let-7*g	Prueitt et al., 2008

1) Focal Adhesion Signaling Pathway

Focal adhesions are large molecular complexes, which are generated following interaction of integrins with Extracellular matrix (ECM). Some of the constituents of focal adhesions are signaling molecules, including different protein kinases and phosphatases, their substrates, and various adapter proteins. Focal adhesion plays an important role in cell signaling. Integrin signaling is dependent upon the non-receptor tyrosine kinase activities of the FAK (Focal adhesion kinase) and src proteins as well as the adaptor protein functions of FAK, V-src sarcoma (Schmidt-Ruppin A-2) viral oncogene (SRC) and SRC homology 2 domain containing) transforming protein 1 (SHC) to initiate downstream signaling events, and plays essential roles in important biological processes including cell motility, cell proliferation, differentiation, and apoptosis (Figure 1). Loss of beta-catenin may result in the disruption of the function of the cell-cell adhesion complex, which may cause weak cell-cell adhesion and confer invasive properties on a tumor (Huiping et al., 2001). Integrin beta 3 is known to be upregulated and plays an important role in melanoma progression and invasion. *Let-7*a reduces invasive potential of melanoma cells by directly repressing expression of integrin beta 3 (Selbach M et al., 2008; Müller and Bosserhoff, 2008). Recent study demonstrated that *Let-7*a decreased the phosphorylation of the AKT and ERK in A549 cells (human lung cancer cell line) (Zhong et al., 2010). The loss of function of *Let-7* resulting in dysregulation of focal adhesion signaling, may be involved in development and progression of cancers

2) Cell Cycle

Mitotic cell cycle progression is accomplished through a reproducible sequence of events, DNA replication (S phase) and mitosis (M phase) separate temporally by gaps known as G1 and G2 phases. Eukaryotic cells respond to DNA damage by activating signaling pathways that promote cell cycle arrest and DNA repair. In this progress, the checkpoint kinase ATM phosphorylates and activates CHK2, which in turn directly phosphorylates and activates p53. p53 and its transcriptional targets play an important role in both G1 and G2 checkpoints. Cyclin-dependent kinases (CDKs) are a family of protein kinases first discovered for their role in regulating the cell cycle. Downstream targets of CDKs include transcription factor E2F and its regulator Rb. Precise activation and inactivation of CDKs at specific points in the cell cycle are required for orderly cell division. Cyclin-CDK inhibitors (CKIs), such as p16INK4A, p15INK4B, p27KIP1, and p21CIP1, are involved in the negative regulation of CDK activities, thus providing a pathway through which the cell cycle is negatively regulated.

Growing evidence indicates that *Let-7* may serve as new negative regulator of cell cycle, suggesting that *Let-7* may be a novel therapeutic candidate for cancer. As shown in Figure 2, multiple signaling components of cell cycle are regulated by *Let-7*. *Let-7* directly regulates a few key cell cycle proto-oncogenes, e.g. CDC25A, CDK6, and cyclin D, thus controlling cell proliferation by reducing flux through the pathways promoting the G1 to S transition (Johnson et al., 2007). In prostate cancer, *Let-7*a induces cell cycle arrest at the G1/S phase by directly targeting E2F2 and CCND2 (Dong et al., 2010).

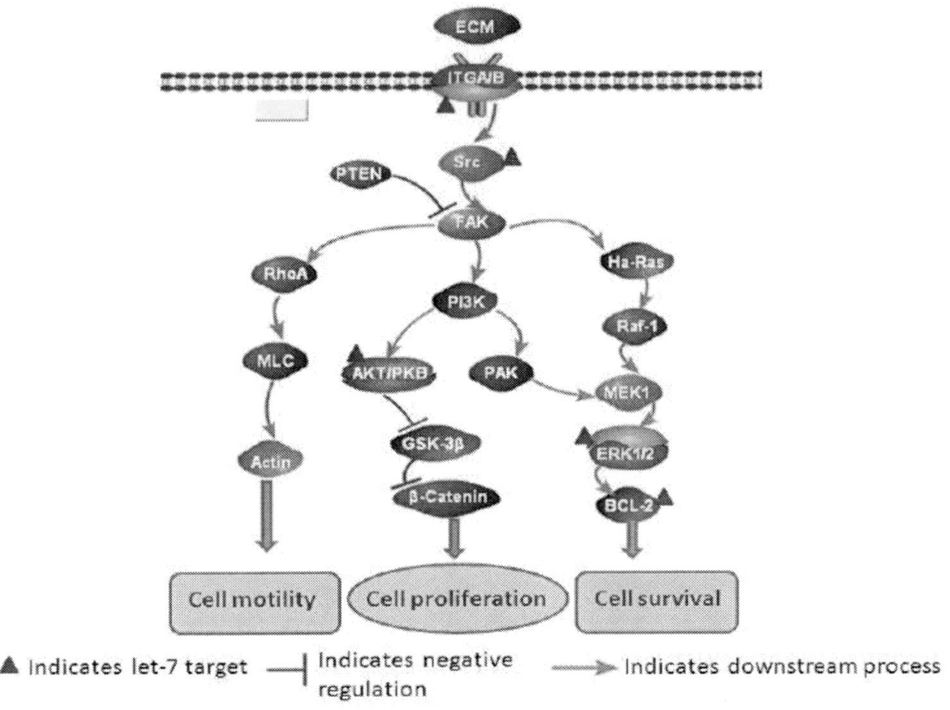

Figure 1. *Let-7* targets several components of focal adhesion signaling pathway. For details see text.

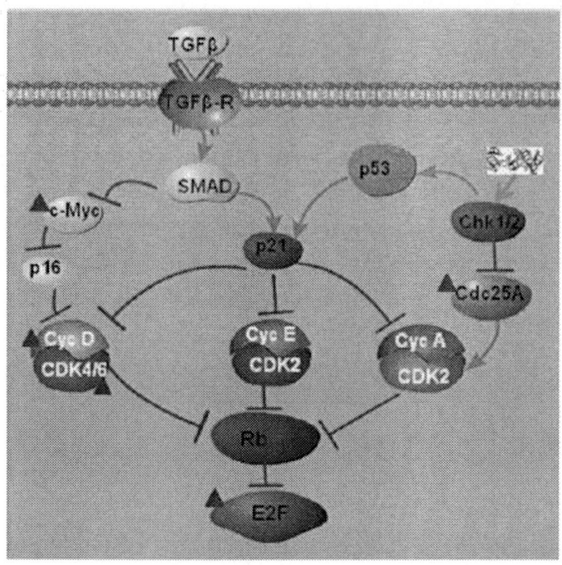

▲ Indicates let-7 target —| Indicated negative regulation —▶ Indicates downstream process

Figure 2. *Let-7* targets various key components of cell cycle signaling pathway. For details see text.

*Let-7*b is shown to cause G2/M arrest through targeting CDC34 resulting in the stabilization of the Wee1 kinase (Legesse-Miller et al., 2009). *Let-7* has also been identified as a tumor suppressor gene that inhibits proliferation in hepatocellular carcinoma (HCC) and Burkitt lymphoma cells by downregulating the oncogene, c-MYC (Sampson et al., 2007; Lan et al., 2011). Recently, an interesting phenomena was observed that repression of *Let-7* targets did not rely solely on the activity of the miRNA alone, as c-MYC repressed in an interdependent manner by *Let-7*b/c and the RNA-binding protein Human antigen R (HuR) (Kim et al., 2009). This additional layer of regulation to miRNA targeting mediated by RNA-binding protein may be a general phenomenon.

3) Apoptosis Signaling Pathway

Apoptosis is a genetically controlled mechanism of cell death involved in the regulation of tissue homeostasis. Apoptosis signaling pathway is always triggered by death receptor engagement, which initiates a signaling cascade mediated by caspase 8 activation. Caspase 8 both feeds directly into caspase 3 activation, which leads to the degradation of cellular proteins necessary to maintain cell survival and integrity (Figure 3). *Let-7* miRNAs have been identified to regulate both pro-apoptotic and anti-apoptotic members. *Let-7* may function as anti-apoptotic gene by targeting FAS and caspase 3. *Let-7* regulates FAS and inhibits FAS-mediated apoptosis in T-cells (Wang et al., 2011). Recent report supports this concept by suggesting that *Let-7* family members regulate endogenous FAS expression in HT29 cells, a line of moderately well-differentiated, non-transformed cells derived from a colon carcinoma (Geng et al., 2011). Previous study has also suggested that *Let-7* expression could influence the differentiation state of tumor cells and their sensitivity to FASL and chemotherapeutic drugs (Shell et al., 2007). Another study indicates that *Let-7*a may play a functional role in modulating drug-induced cell death in human cancer cells by targeting caspase 3 (Tsang and Kwok, 2008).

However, *Let-7* miRNAs negatively regulate BCL-xL, an anti-apoptotic member of the BCL2, and induce apoptosis in cooperation with an anti-cancer drug targeting MCL1 in human hepatocellular carcinomas (Shimizu et al., 2010). *Let-7* miRNAs have different function in the same physiological or pathological conditions, which may be due to regulation of different targets in different types of cells, or in different cell stages.

4) MAPK

The mitogen-activated protein kinase (MAPK) cascade is a highly conserved module that is involved in various cellular functions, including cell proliferation, differentiation and migration. Cytokines, growth factors and Toll-like receptor (TLR) ligands can initiate MAPK pathway and regulate at least four signal transduction processes, such as ERK, JNK, p38 and NFKB (Figure 4). The TLR pathways are key regulators in cancer progression as well as chemoresistance (Chen et al., 2008). *Let-7* miRNAs regulate Toll-like receptor 4 in immune system (Chen et al., 2007; Androulidaki et al., 2009). *Let-7* involves in RAS/MEK pathway by direct regulating Ras expression (Kumar et al., 2008). In pancreatic cancer cells, the knockdown of HMGA2, as a *Let-7* target, causes epithelial-mesenchymal transition (EMT) and growth suppression similar to inhibition of the RAS/MEK pathway (Watanabe et al., 2009). Inhibition of MAPK by Raf kinase inhibitory protein (RKIP) elevates *Let-7* expression, resulting in downregulation of HMGA2 and suppression of invasion and metastasis in breast cancer cells (Dangi-Garimella et al., 2009).

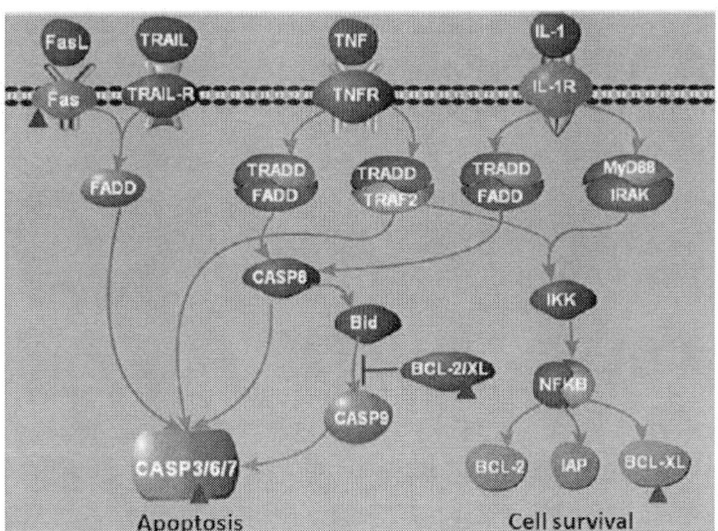

Figure 3. *Let-7* targets several components of cell apoptosis signaling pathway. For details see text.

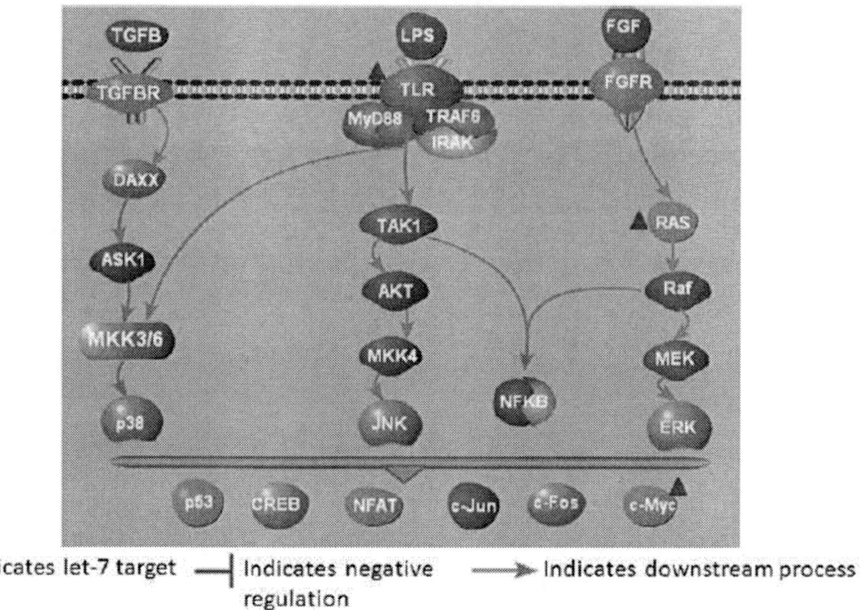

Figure 4. *Let-7* targets several key components of MAPK signaling pathway. For details see text.

5. Single-Nucleotide Polymorphisms (SNPs) of *Let-7* Binding Sites and Cancer

It has become emerging concept that the deregulation of miRNAs participates in the activation of cell proliferation or inactivation of apoptotic signaling pathway in conjunction with genetic changes leading to cancer pathogenesis. Single nucleotide polymorphisms (SNPs) located in *Let-7* binding sites may perturb miRNA function by introducing or disrupting binding sites. A SNP in the KRAS 3' UTR disrupts *Let-7* binding site, resulting in KRAS overexpression, which increases non-small cell lung cancer risk and is associated with reduced survival in oral cancers (Chin et al., 2008; Christensen et al., 2009). Motivated by this concept, we browsed the database: Polymorphism in microRNA Target Site (PolymiRTS), a database of naturally occurring DNA variations in putative microRNA (miRNA) target sites (Bao et al., 2007), and found that 31 SNPs located in *Let-7* binding sites (Table 3). It is necessary to determine whether these SNPs affect *Let-7* function and are associated with human diseases.

6. Regulation of *Let-7* in Signaling Feedback Loop

Like other miRNAs, *Let-7* originates from large primary (pri) and precursor (pre) transcript that undergoes various processing steps along its biogenesis pathway till it reaches its mature and functional form (Davis-Dusenbery and Hata, 2010). The *Let-7* family members are initially transcribed by RNA polymerase II from the genome as long primary transcripts (pri-*Let-7*). These transcripts are first processed to precursor (pre-*Let-7*) by the microprocessor complex containing the RNase-III enzyme Drosha and the double-stranded RNA-binding protein Pasha (DGCR8). The pre-*Let-7* is exported from the nucleus to the cytoplasm by a Ran-GTP transporter known as exportin 5. In the cytoplasm, the pre-*Let-7* is further processed by a second RNase-III enzyme known as Dicer (which complexes with TRBP and AGO2) to generate mature *Let-7*. Regulators that affect the biogenesis of *Let-7* may be at transcriptional level, including transcription factors and epigenetic modifications, and at posttranscriptional level, such as some proteins that affect the activity of Drosha or Dicer (Figure 5).

Table 3. SNP located in *Let-7* binding sites

Transcipt ID	Chromosome	SNPID	Allele1	Allele2	Allele 1 binding miRNAs	Allele 2 binding miRNAs
NM_014936	6	rs1048072	A	T	miR-196a, miR-196b,*Let-7*	
BX647816	16	rs1051758	C	T	*Let-7*	
NM_014352	11	rs11217815	A	G	*Let-7*,miR-202	
NM_144701	1	rs11465828	A	G	*Let-7*	
NM_138983	21	rs11554599	C	T	*Let-7*	
NM_001001930	22	rs11703765	C	T	*Let-7*,miR-202	
NM_001066	1	rs11807940	C	T	*Let-7*,miR-202	
NM_198526	15	rs12148341	A	T	*Let-7*	
NM_001001678	10	rs12783098	T	A	*Let-7*,miR-202	
AK126701	3	rs13058936	T	C	miR-335,*Let-7*	
AK021848	6	rs16882165	A	C	*Let-7*,miR-202	
NM_004253	9	rs16910873	C	A	*Let-7*,miR-202	
NM_015455	5	rs17080395	C	T	*Let-7*a,miR-202	
NM_001025579	17	rs17852347	A	G	*Let-7*	miR-103, miR-107,
NM_014480	19	rs187760	A	G	*Let-7*,miR-202	
NM_080282	17	rs2233766	T	C	*Let-7*,miR-202	
NM_004342	7	rs2718145	C	T	*Let-7*,miR-202	
AB046770	7	rs281896	T	C	*Let-7*	miR-620,
NM_015488	2	rs3204750	C	T	*Let-7*	miR-19a,miR-19b,miR-527
NM_145042	22	rs383094	C	A	miR-196a,miR-196b,*Let-7*	miR-143
AY341951	12	rs4018436	T	C	*Let-7*	
NM_022901	9	rs4993020	T	C	*Let-7*	
NM_181443	20	rs6104808	C	T	*Let-7*,miR-202	
NM_018498	20	rs614507	C	A	*Let-7*,miR-202	
NM_013434	2	rs6730587	A	G	*Let-7*	
NM_006262	12	rs7137584	A	C	*Let-7*	
NM_017728	17	rs7212248	A	G	*Let-7*	
AK055428	2	rs7607457	C	T	*Let-7*,miR-202	
AK130265	17	rs8081375	A	G	*Let-7*	
NM_001005354	12	rs8707	C	G	*Let-7*	miR-657
NM_004391	3	rs9816972	T	C	*Let-7*,miR-202	miR-625

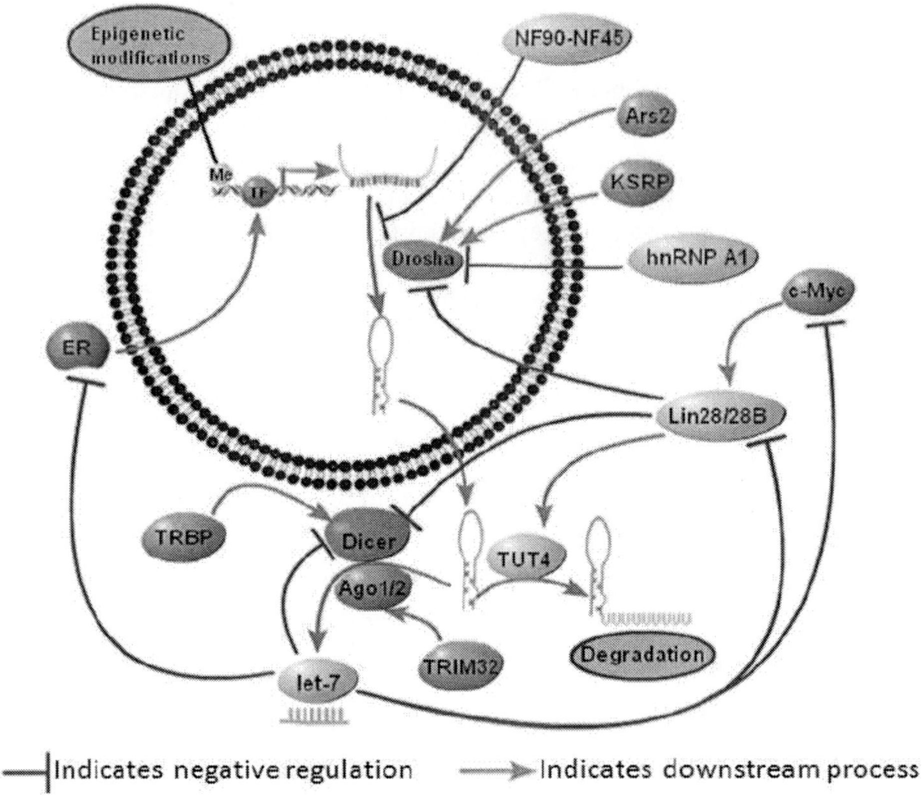

Figure 5. Regulation of the biogenesis of *Let-7* by transcriptianal and posttranscriptional processes. For details see text.

Epigenetics plays great role in gene expression regulation. It has been shown that miRNA expression also can be regulated by DNA methylation. *Let-7*a-3 hypomethylation facilitates epigenetic reactivation of the gene and elevates expression of *Let-7*a-3 in a human lung cancer cell line (Brueckner et al., 2007). Transforming growth factor-beta down-regulates *Let-7*d expression through SMAD3 binding to the *Let-7*d promoter (Pandit et al., 2010). *Let-7* biogenesis is also highly regulated process involving multiple proteins at various stages. A complex of nuclear factor 90 (NF90) and NF45 proteins functions as a negative regulator in miRNA biogenesis due to impairing access of the Microprocessor complex to the pri-*Let-7*a (Sakamoto et al., 2009). The KH-type splicing regulatory protein, KSRP, can serve as a component of both Drosha and Dicer complexes and regulates the biogenesis of a subset of miRNAs including *Let-7*. KSRP binds with high affinity to the terminal loop of the miRNA precursors and promotes their maturation (Trabucchi et al., 2009). Heteronuclear ribonucleoprotein A1 (hnRNP A1) binding to the conserved terminal loop of pri-*Let-7*a interferes the binding of KSRP and inhibits the *Let-7*a processing by Drosha (Michlewski and Cáceres, 2010). Arsenic resistance protein 2 (Ars2), a component of the nuclear RNA cap-binding complex (CBC), is also important for the biogenesis of several miRNAs including miR-21, *Let-7* and miR-155 (Gruber et al., 2009).

LIN28/28B, as a highly conserved RNA-binding protein, has been identified that regulates the processing of *Let-7* miRNA. LIN28/28B represses *Let-7* expression by three mechanisms identified by different groups. LIN28/28B inhibits *Let-7* processing by Drosha or by Dicer. LIN28 can also recruit TUT4, as the uridylyl transferase, to pre-*Let-7* and adds an oligouridine tail to the pre-*Let-7*, which blocks Dicer processing (Boyerinas et al., 2010; Heo et al., 2009). In aggressive liver cancer, LIN28 and LIN28B are upregulated and negatively correlated with expression of several *Let-7* family members, including *Let-7*a (Cairo et al., 2010).

Some regulators of *Let-7* expression are *Let-7* targets, e.g. LIN28/28B and Dicer, which constitutes feedback loops. There is a positive-feedback loop involving NFKB, LIN28B, *Let-7*, and IL6: NFKB induces expression of LIN28B, leading to repression of *Let-7* and expression of the gene encoding IL6, a *Let-7* target (Iliopoulos et al., 2009). The MYC, as a *Let-7* target, upregualtes LIN28 and LIN28B and inhibits biogenesis of *Let-7* family members (Chang et al., 2009). Dicer, known as a mediator of miRNA maturation, is also regulated by *Let-7*, suggesting the existence of a finely tuned regulatory loop (Tokumaru et al., 2008).

Conclusion

As we have discussed throughout this chapter, the *Let-7* family plays a role in an exceedingly diverse array of cellular activities. There is a very clear link between deregulation of *Let-7* and human cancers. *Let-7* is widely viewed as tumor suppressor. Loss of *Let-7* expression is associated with the development of poorly differentiated, aggressive cancers. However, upregulation of certain *Let-7* family members in cancers has also been observed, suggesting that *Let-7* does not play a tumor suppressor function under all circumstances and/or in all tissues. In addition, loss of *Let-7* increases resistance to certain chemotherapeutic drugs. It remains to be answered whether we can use the expression of *Let-7* family only or combined with other cancer-related miRNAs for cancer prognosis and/or selection of chemotherapy.

Let-7 has been shown to be integrated into vast regulatory networks that impinge upon a broad spectrum of biological events. The deregulated *Let-7* and other miRNAs, and their targets construct a complex interacting network contributed to tumorigenesis. *Let-7* functions as tumor suppressor during tumor development and progression. Experimental evidence demonstrates ectopic *Let-7* expression using mimics can normalize the gene regulatory network and signaling pathways, and reverse the phenotype in cancerous cells. Restoration of *Let-7* expression to tumors holds great therapeutic potential for the treatment of these aggressive types of cancer. Questions remain, however, as to whether different family members indeed have specific activities by targeting different genes in a particular cell type.

As we have elucidated, *Let-7* expression is regulated on multiple levels, and several proteins appear to regulate the *Let-7* transcription or processing. Particular *Let-7* family members have been shown to be specifically deregulated in certain cancers. Epigenetic regulation and processing at transcription and post-transcriptional level are probably involved in *Let-7* deregulation in cancers. The precise transcriptional and processing regulation mechanisms need to be determined in the future, which may provide a potential alternative therapeutic strategy.

REFERENCES

Akao, Y., Nakagawa, Y., Naoe, T. *Let-7* microRNA functions as a potential growth suppressor in human colon cancer cells. *Biol. Pharm. Bull.* 2006;29:903-906.

Androulidaki, A., Iliopoulos, D., Arranz, A., Doxaki, C., Schworer, S., Zacharioudaki, V., Margioris, A. N., Tsichlis, P. N., Tsatsanis, C. The kinase Akt1 controls macrophage response to lipopolysaccharide by regulating microRNAs. *Immunity*. 2009;31:220-231.

Bao, L., Zhou, M., Wu, L., Lu, L., Goldowitz, D., Williams, R. W., Cui, Y. PolymiRTS Database: linking polymorphisms in microRNA target sites with complex traits. *Nucleic Acids Res*. 2007;35:D51-54.

Boyerinas, B., Park, S. M., Hau, A., Murmann, A. E., Peter, M. E. The role of *Let-7* in cell differentiation and cancer. *Endocr. Relat. Cancer.* 2010;17:F19-F36.Brueckner, B., Stresemann, C., Kuner, R., Mund, C., Musch, T., Meister, M., Sültmann, H., Lyko, F. The human *Let-7*a-3 locus contains an epigenetically regulated microRNA gene with oncogenic function. *Cancer Res*. 2007;67:1419-1423.

Budhu, A., Jia, H. L., Forgues, M., Liu, C. G., Goldstein, D., Lam, A., Zanetti, K. A., Ye, Q. H., Qin, L. X., Croce, C. M., Tang, Z. Y., Wang, X. W. Identification of metastasis-related microRNAs in hepatocellular carcinoma. *Hepatology*. 2008;47:897-907

Cairo, S., Wang, Y., De Reyniès, A., Duroure, K., Dahan, J., Redon, M. J., Fabre, M., McClelland, M., Wang, X. W., Croce, C. M., Buendia, M. A. Stem cell-like micro-RNA signature driven by Myc in aggressive liver cancer. *Proc. Natl. Acad. Sci.* U S A. 2010;107:20471-20476.

Cammarata, G., Augugliaro, L., Salemi, D., Agueli, C., La Rosa, M., Dagnino, L., Civiletto, G., Messana, F., Marfia, A., Bica, M. G., Cascio, L., Floridia, P. M., Mineo, A. M., Russo, M., Fabbiano, F., Santoro, A. Differential expression of specific microRNA and their targets in acute myeloid leukemia. *Am. J. Hematol*. 2010;85:331-339.

Chang, T. C., Zeitels, L. R., Hwang, H. W., Chivukula, R. R., Wentzel, E. A., Dews, M., Jung, J., Gao, P., Dang, C. V., Beer, M. A., Thomas-Tikhonenko, A., Mendell, J. T. LIN28B transactivation is necessary for Myc-mediated *Let-7* repression and proliferation. *Proc. Natl. Acad. Sci.*U S A. 2009;106: 3384-3389.

Chen, R., Alvero, A. B., Silasi, D. A., Steffensen, K. D., Mor, G. Cancers take their Toll--the function and regulation of Toll-like receptors in cancer cells. *Oncogene*. 2008;27:225-233

Chen, X., Ba, Y., Ma, L., Cai, X., Yin, Y., Wang, K., Guo, J., Zhang, Y., Chen, J., Guo, X., Li, Q., Li, X., Wang, W., Zhang, Y., Wang, J., Jiang, X., Xiang, Y., Xu, C., Zheng, P., Zhang, J., Li, R., Zhang, H., Shang, X., Gong, T., Ning, G., Wang, J., Zen, K., Zhang, J., Zhang, C. Y. Characterization of microRNAs in serum: a novel class of biomarkers for diagnosis of cancer and other diseases. *Cell Res*. 2008; 18:997-1006.

Chen, X. M., Splinter, P. L., O'Hara, S. P., LaRusso, N. F. A cellular micro-RNA, *Let-7*i, regulates Toll-like receptor 4 expression and contributes to cholangiocyte immune responses against Cryptosporidium parvum infection. *J. Biol. Chem.* 2007;282:28929-28938.

Chen, Y., Stallings, R. L. Differential patterns of microRNA expression in neuroblastoma are correlated with prognosis, differentiation, and apoptosis. *Cancer Res*. 2007;67:976-983.

Chin, L. J., Ratner, E., Leng, S., Zhai, R., Nallur, S., Babar, I., Muller, R. U., Straka, E., Su, L., Burki, E. A., Crowell, R. E., Patel, R., Kulkarni, T., Homer, R., Zelterman, D., Kidd,

K. K., Zhu, Y., Christiani, D. C., Belinsky, S. A., Slack, F. J., Weidhaas, J. B. A SNP in a *Let-7* microRNA complementary site in the KRAS 3' untranslated region increases non-small cell lung cancer risk. *Cancer Res.* 2008; 68:8535-8540.

Christensen, B. C., Moyer, B. J., Avissar, M., Ouellet, L. G., Plaza, S. L., McClean, M. D., Marsit, C. J., Kelsey, K. T. A *Let-7* microRNA-binding site polymorphism in the KRAS 3' UTR is associated with reduced survival in oral cancers. *Carcinogenesis.* 2009;30:1003-1007.

Cummins, J. M., He, Y., Leary, R. J., Pagliarini, R., Diaz, L. A. Jr, Sjoblom, T., Barad, O., Bentwich, Z., Szafranska, A. E., Labourier, E., Raymond, C. K., Roberts, B. S., Juhl, H., Kinzler, K. W., Vogelstein, B., Velculescu, V. E. The colorectal microRNAome. *Proc. Natl. Acad. Sci.* U S A. 2006;103: 3687-3692.

Dahiya, N., Sherman-Baust, C. A., Wang, T. L., Davidson, B., Shih, IeM, Zhang, Y., Wood, W. 3rd, Becker, K. G., Morin, P. J. MicroRNA expression and identification of putative miRNA targets in ovarian cancer. *PLoS One.* 2008;3:e2436.

Dangi-Garimella, S., Yun, J., Eves, E. M., Newman, M., Erkeland, S. J., Hammond, S. M., Minn, A. J., Rosner, M. R. Raf kinase inhibitory protein suppresses a metastasis signaling cascade involving LIN28 and *Let-7*. *EMBO J.* 2009;28:347-358.

Davis-Dusenbery BN, Hata A. Mechanisms of control of microRNA biogenesis. J Biochem. 2010 Oct;148(4):381-92.

Dixon-McIver, A., East, P., Mein, C. A., Cazier, J. B., Molloy, G., Chaplin, T., Andrew Lister, T., Young, B. D., Debernardi, S. Distinctive patterns of microRNA expression associated with karyotype in acute myeloid leukaemia. *PLoS One.* 2008;3:e2141.

Dong, Q., Meng, P., Wang, T., Qin, W., Qin, W., Wang, F., Yuan, J., Chen, Z., Yang, A., Wang, H. MicroRNA *Let-7*a inhibits proliferation of human prostate cancer cells in vitro and in vivo by targeting E2F2 and CCND2. *PLoS One.* 2010;5:e10147.

Du Rieu, M. C., Torrisani, J., Selves, J., Al Saati, T., Souque, A., Dufresne, M., Tsongalis, G. J., Suriawinata, A. A., Carrère, N., Buscail, L., Cordelier, P. MicroRNA-21 is induced early in pancreatic ductal adenocarcinoma precursor lesions. *Clin. Chem.* 2010;56:603-612.

Dweep, H., Sticht, C., Pandey, P., Gretz, N. miRWalk - database: prediction of possible miRNA binding sites by "walking" the genes of 3 genomes. *J. Biomed. Inform.* 2011;44:839-847.

Fang, W. J., Lin, C. Z., Zhang, H. H., Qian, J., Zhong, L., Xu, N. Detection of *Let-7*a microRNA by real-time PCR in colorectal cancer: a single-centre experience from China. *J. Int. Med. Res.* 2007;35:716-723.

Garzon, R., Garofalo, M., Martelli, M. P., Briesewitz, R., Wang, L., Fernandez-Cymering, C., Volinia, S., Liu, C. G., Schnittger, S., Haferlach, T., Liso, A., Diverio, D., Mancini, M., Meloni, G., Foa, R., Martelli, M. F., Mecucci, C., Croce, C. M., Falini, B. Distinctive microRNA signature of acute myeloid leukemia bearing cytoplasmic mutated nucleophosmin. *Proc. Natl. Acad. Sci.* U S A. 2008;105: 3945-3950.

Geng, L., Zhu, B., Dai, B. H., Sui, C. J., Xu, F., Kan, T., Shen, W. F., Yang, J. M. A *Let-7*/Fas double-negative feedback loop regulates human colon carcinoma cells sensitivity to Fas-related apoptosis. *Biochem. Biophys. Res. Commun.* 2011;408:494-499.

Gramantieri, L., Ferracin, M., Fornari, F., Veronese, A., Sabbioni, S., Liu, C. G., Calin, G. A., Giovannini, C., Ferrazzi, E., Grazi, G. L., Croce, C. M., Bolondi, L., Negrini, M. Cyclin

G1 is a target of miR-122a, a microRNA frequently down-regulated in human hepatocellular carcinoma. *Cancer Res*. 2007;67: 6092-6099.

Gruber, J. J., Zatechka, D. S., Sabin, L. R., Yong, J., Lum, J. J., Kong, M., Zong, W. X., Zhang, Z., Lau, C. K., Rawlings, J., Cherry, S., Ihle, J. N., Dreyfuss, G., Thompson, C. B. Ars2 links the nuclear cap-binding complex to RNA interference and cell proliferation. *Cell*. 2009;138:328-339.

Hagen, J. W., Lai, E. C. microRNA control of cell-cell signaling during development and disease. *Cell Cycle*. 2008;7:2327-2332.

Heo, I., Joo, C., Kim, Y. K., Ha, M., Yoon, M. J., Cho, J., Yeom, K. H., Han, J., Kim, V. N. TUT4 in concert with LIN28 suppresses microRNA biogenesis through pre-microRNA uridylation. *Cell*. 2009;138:696-708.

Heo, I., Kim, V. N. Regulating the regulators: posttranslational modifications of RNA silencing factors. *Cell*. 2009;139:28-31.

Huang, Y. S., Dai, Y., Yu, X. F., Bao, S. Y., Yin, Y. B.,Tang, M., Hu, C. X. Microarray analysis of microRNA expression in hepatocellular carcinoma and non-tumorous tissues without viral hepatitis. *J. Gastroenterol. Hepatol*. 2008;23:87-94.

Huiping, C., Kristjansdottir, S., Jonasson, J. G., Magnusson, J., Egilsson, V., Ingvarsson, S. Alterations of E-cadherin and beta-catenin in gastric cancer. *BMC Cancer*.2001;1:16.

Iliopoulos, D., Hirsch, H. A., Struhl, K. An epigenetic switch involving NF-kappaB, LIN28, *Let-7* MicroRNA, and IL6 links inflammation to cell transformation. *Cell*. 2009;139:693-706.

Iorio, M. V., Ferracin, M., Liu, C. G., Veronese, A., Spizzo, R., Sabbioni, S., Magri, E., Pedriali, M., Fabbri, M., Campiglio, M., Ménard, S., Palazzo, J. P., Rosenberg, A., Musiani, P., Volinia, S., Nenci, I., Calin, G. A., Querzoli, P., Negrini, M., Croce, C. M. MicroRNA gene expression deregulation in human breast cancer. *Cancer Res*. 2005;65:7065-7070.

Iorio, M. V., Visone, R., Di Leva, G., Donati, V., Petrocca, F., Casalini, P., Taccioli, C., Volinia, S., Liu, C. G., Alder, H., Calin, G. A., Ménard, S., Croce, C. M. MicroRNA signatures in human ovarian cancer. *Cancer Res*. 2007;67:8699-8707.

Jiang, J., Gusev, Y., Aderca, I., Mettler, T. A., Nagorney, D. M., Brackett, D. J., Roberts, L. R., Schmittgen, T. D. Association of MicroRNA expression in hepatocellular carcinomas with hepatitis infection, cirrhosis, and patient survival. *Clin. Cancer Res*. 2008;14:419-427.

Johnson, C. D., Esquela-Kerscher, A., Stefani, G., Byrom, M., Kelnar, K., Ovcharenko, D., Wilson, M., Wang, X., Shelton, J., Shingara, J., Chin, L., Brown, D., Slack, F. J. The *Let-7* microRNA represses cell proliferation pathways in human cells. *Cancer Res*. 2007;67:7713-7722.

Johnson, S. M., Grosshans, H., Shingara, J., Byrom, M., Jarvis, R., Cheng, A., Labourier, E., Reinert, K. L., Brown, D., Slack, F. J. RAS is regulated by the *Let-7* microRNA family. *Cell*. 2005;120:635-647.

Jongen-Lavrencic, M., Sun, S. M., Dijkstra, M. K., Valk, P. J., Löwenberg, B. MicroRNA expression profiling in relation to the genetic heterogeneity of acute myeloid leukemia. *Blood*. 2008;111: 5078-5085.

Kim, H. H., Kuwano, Y., Srikantan, S., Lee, E. K., Martindale, J. L., Gorospe, M. HuR recruits *Let-7*/RISC to repress c-Myc expression. *Genes Dev*. 2009;23:1743-1748.

Kumar, M. S., Erkeland, S. J., Pester, R. E., Chen, C. Y., Ebert, M. S., Sharp, P. A., Jacks, T. Suppression of non-small cell lung tumor development by the *Let-7* microRNA family. *Proc. Natl. Acad. Sci.* U S A. 2008;105:3903-3908.

Laios, A., O'Toole, S., Flavin, R., Martin, C., Kelly, L., Ring, M., Finn, S.P., Barrett, C., Loda, M., Gleeson, N., D'Arcy, T., McGuinness, E., Sheils, O., Sheppard, B., O' Leary, J. Potential role of miR-9 and miR-223 in recurrent ovarian cancer. *Mol. Cancer.* 2008;7:35.

Lan, F. F., Wang, H., Chen, Y. C., Chan, C. Y., Ng, S. S., Li, K., Xie, D., He, M. L., Lin, M. C., Kung, H.F. Hsa-*Let-7*g inhibits proliferation of hepatocellular carcinoma cells by downregulation of c-Myc and upregulation of p16(INK4A) *Int. J. Cancer.* 2011;128:319-331.

Lawrie, C. H., Chi, J., Taylor, S., Tramonti, D., Ballabio, E., Palazzo, S., Saunders, N. J., Pezzella, F., Boultwood, J., Wainscoat, J.S., Hatton, C. S. Expression of microRNAs in diffuse large B cell lymphoma is associated with immunophenotype, survival and transformation from follicular lymphoma. *J. Cell Mol. Med.* 2009;13:1248-1260.

Lee, E. J., Gusev, Y., Jiang, J., Nuovo, G. J., Lerner, M. R., Frankel, W. L., Morgan, D. L., Postier, R. G., Brackett, D. J., Schmittgen, T. D. Expression profiling identifies microRNA signature in pancreatic cancer. *Int. J. Cancer.* 2007;120:1046-1054.

Legesse-Miller, A., Elemento, O., Pfau, S. J., Forman, J. J., Tavazoie, S., Coller, H. A. *J. Biol. Chem. Let-7* Overexpression leads to an increased fraction of cells in G2/M, direct down-regulation of Cdc34, and stabilization of Wee1 kinase in primary fibroblasts. 2009;284:6605-6609.

Liu, C. C., Lin, C. C., Chen, W. S., Chen, H. Y., Chang, P. C., Chen, J. J., Yang, P. C. CRSD: a comprehensive web server for composite regulatory signature discovery. *Nucleic Acids Res.* 2006;34:W571-W577.

Michlewski, G., Cáceres, J. F. Antagonistic role of hnRNP A1 and KSRP in the regulation of *Let-7*a biogenesis. *Nat. Struct. Mol. Biol.* 2010;17:1011-1018.

Monzo, M., Navarro, A., Bandres, E., Artells, R., Moreno, I., Gel, B., Ibeas, R., Moreno, J., Martinez, F., Diaz, T., Martinez, A., Balagué, O., Garcia-Foncillas, J. Overlapping expression of microRNAs in human embryonic colon and colorectal cancer. *Cell Res.* 2008:18:823-833.

Müller, D. W., Bosserhoff, A. K. Integrin beta 3 expression is regulated by *Let-7*a miRNA in malignant melanoma. *Oncogene.* 2008;27:6698-6706.

Nakajima, G., Hayashi, K., Xi, Y., Kudo, K., Uchida, K., Takasaki, K., Yamamoto, M., Ju, J. Non-coding MicroRNAs hsa-*Let-7*g and hsa-miR-181b are Associated with Chemoresponse to S-1 in Colon Cancer. *Cancer Genomics Proteomics.* 2006;3:317-324.

Nam, E. J., Yoon, H., Kim, S. W., Kim, H., Kim, Y. T., Kim, J. H., Kim, J. W., Kim, S. MicroRNA expression profiles in serous ovarian carcinoma. *Clin. Cancer Res.* 2008;14:2690-2695.

Ohshima, K., Inoue, K., Fujiwara, A., Hatakeyama, K., Kanto, K., Watanabe, Y., Muramatsu, K., Fukuda, Y., Ogura, S., Yamaguchi, K., Mochizuki, T. *Let-7* microRNA family is selectively secreted into the extracellular environment via exosomes in a metastatic gastric cancer cell line. *PLoS One.* 2010;5:e13247.

Pandit, K. V., Corcoran, D., Yousef, H., Yarlagadda, M., Tzouvelekis, A., Gibson, K. F., Konishi, K., Yousem, S. A., Singh, M., Handley, D., Richards, T., Selman, M., Watkins, S. C., Pardo, A., Ben-Yehudah, A., Bouros, D., Eickelberg, O., Ray, P., Benos, P. V.,

Kaminski, N. Inhibition and role of *Let-7*d in idiopathic pulmonary fibrosis. *Am. J. Respir. Crit. Care Med.* 2010;182:220-229.

Peltier, H. J., Latham, G. J. Normalization of microRNA expression levels in quantitative RT-PCR assays: identification of suitable reference RNA targets in normal and cancerous human solid tissues. *RNA*. 2008;14:844-852.

Pineau, P., Volinia, S., McJunkin, K., Marchio, A., Battiston, C., Terris, B., Mazzaferro, V., Lowe, S. W., Croce, C. M., Dejean, A. miR-221 overexpression contributes to liver tumorigenesis. *Proc. Natl. Acad. Sci.* U S A. 2010;107:264-269.

Porkka, K. P., Pfeiffer, M. J., Waltering, K. K., Vessella, R. L., Tammela, T. L., Visakorpi, T. MicroRNA expression profiling in prostate cancer. *Cancer Res.* 2007;67:6130-6135.

Prueitt, R. L., Yi, M., Hudson, R. S., Wallace, T. A., Howe, T. M., Yfantis, H. G., Lee, D. H., Stephens, R. M., Liu, C. G., Calin, G. A., Croce, C. M., Ambs, S. Expression of microRNAs and protein-coding genes associated with perineural invasion in prostate cancer. *Prostate*. 2008;68:1152-1164.

Ruepp A, Kowarsch A, Schmidl D, Buggenthin F, Brauner B, Dunger I, Fobo G, Frishman G, Montrone C, Theis FJ. PhenomiR: a knowledgebase for microRNA expression in diseases and biological processes. Genome Biology 2010, 11:R6

Sampson, V. B., Rong, N. H., Han, J., Yang, Q., Aris, V., Soteropoulos, P., Petrelli, N. J., Dunn, S. P., Krueger, L. J. MicroRNA *Let-7*a down-regulates MYC and reverts MYC-induced growth in Burkitt lymphoma cells. *Cancer Res*. 2007;67:9762-9770.

Sakamoto, S., Aoki, K., Higuchi, T., Todaka, H., Morisawa, K., Tamaki, N., Hatano, E., Fukushima, A., Taniguchi, T., Agata, Y. The NF90-NF45 complex functions as a negative regulator in the microRNA processing pathway. *Mol. Cell Biol.* 2009;29:3754-3769.

Schulte, J. H., Horn, S., Otto, T., Samans, B., Heukamp, L. C., Eilers, U. C., Krause, M., Astrahantseff, K., Klein-Hitpass, L., Buettner, R., Schramm, A., Christiansen, H., Eilers, M., Eggert, A., Berwanger, B. MYCN regulates oncogenic MicroRNAs in neuroblastoma. *Int. J. Cancer*. 2008;122:699-704.

Schwamborn, J. C., Berezikov, E., Knoblich, J. A. The TRIM-NHL protein TRIM32 activates microRNAs and prevents self-renewal in mouse neural progenitors. *Cell.* 2009;136:913-925.

Selbach, M., Schwanhäusser, B., Thierfelder, N., Fang, Z., Khanin, R., Rajewsky, N. Widespread changes in protein synthesis induced by microRNAs. *Nature*. 2008;455:58-63.

Shell, S., Park, S. M., Radjabi, A. R., Schickel, R., Kistner, E. O., Jewell, D. A., Feig, C., Lengyel, E., Peter, M. E. *Let-7* expression defines two differentiation stages of cancer. *Proc. Natl. Acad. Sci.* U S A.2007;104:11400-11405.

Sempere, L. F., Christensen, M., Silahtaroglu, A., Bak, M., Heath, C. V., Schwartz, G., Wells, W., Kauppinen, S., Cole, C. N. Altered MicroRNA expression confined to specific epithelial cell subpopulations in breast cancer. *Cancer Res*. 2007;67:11612-11622.

Shimizu, S., Takehara, T., Hikita, H., Kodama, T., Miyagi, T., Hosui, A., Tatsumi, T., Ishida, H., Noda, T., Nagano, H., Doki, Y., Mori, M., Hayashi, N. The *Let-7* family of microRNAs inhibits Bcl-xL expression and potentiates sorafenib-induced apoptosis in human hepatocellular carcinoma. *J. Hepatol.* 2010;52:698-704.

Takamizawa, J., Konishi, H., Yanagisawa, K., Tomida, S., Osada, H., Endoh, H., Harano, T., Yatabe, Y., Nagino, M., Nimura, Y., Mitsudomi, T., Takahashi, T. Reduced expression of the *Let-7* microRNAs in human lung cancers in association with shortened postoperative survival. *Cancer Res*. 2004;64:3753-3756.

Tokumaru, S., Suzuki, M., Yamada, H., Nagino, M., Takahashi, T. *Let-7* regulates Dicer expression and constitutes a negative feedback loop. *Carcinogenesis*. 2008;29:2073-2077.

Trabucchi, M., Briata, P., Garcia-Mayoral, M., Haase, A. D., Filipowicz, W., Ramos, A., Gherzi, R., Rosenfeld, M. G. The RNA-binding protein KSRP promotes the biogenesis of a subset of microRNAs. *Nature*. 2009;459:1010-1014.

Tsang, W. P., Kwok, T. T. *Let-7*a microRNA suppresses therapeutics-induced cancer cell death by targeting caspase-3. *Apoptosis*. 2008;13: 215-1222.

Varnholt, H., Drebber, U., Schulze, F., Wedemeyer, I., Schirmacher, P., Dienes, H. P., Odenthal, M. MicroRNA gene expression profile of hepatitis C virus-associated hepatocellular carcinoma. *Hepatology*. 2008;47:1223-1232.

Volinia, S., Calin, G. A., Liu, C. G., Ambs, S., Cimmino, A., Petrocca, F., Visone, R., Iorio, M., Roldo, C., Ferracin, M., Prueitt, R. L., Yanaihara, N., Lanza, G., Scarpa, A., Vecchione, A., Negrini, M., Harris, C. C., Croce, C. M. A microRNA expression signature of human solid tumors defines cancer gene targets. *Proc. Natl. Acad. Sci.* U S A. 2006;103:2257-2261.

Wang, E. MicroRNA systems biology. In: RNA Technologies in Cardiovascular Medicine and Research. Wolfgang, P. and Jan, B. Eds. *Springer*, 2008;69-86.

Wang, S., Tang, Y., Cui, H., Zhao, X., Luo, X., Pan, W., Huang, X., Shen, N. *Let-7*/miR-98 regulate Fas and Fas-mediated apoptosis. *Genes Immun.* 2011;12:149-154.

Watanabe, S., Ueda, Y., Akaboshi, S., Hino, Y., Sekita, Y., Nakao, M. HMGA2 maintains oncogenic RAS-induced epithelial-mesenchymal transition in human pancreatic cancer cells. *Am. J. Pathol.* 2009;174:854-868.

Yanaihara, N., Caplen, N., Bowman, E., Seike, M., Kumamoto, K., Yi, M., Stephens, R. M., Okamoto, A., Yokota, J., Tanaka, T., Calin, G. A., Liu, C. G., Croce, C. M., Harris, C. C. Unique microRNA molecular profiles in lung cancer diagnosis and prognosis. *Cancer Cell.* 2006;9:189-198.

Yang, H., Kong, W., He, L., Zhao, J. J., O'Donnell, J. D., Wang, J., Wenham, R. M., Coppola, D., Kruk, P. A., Nicosia, S. V., Cheng, J. Q. MicroRNA expression profiling in human ovarian cancer: miR-214 induces cell survival and cisplatin resistance by targeting PTEN. *Cancer Res*. 2008;68:425-433.

Zhang, H. H., Wang, X. J., Li, G. X., Yang, E., Yang, N. M. Detection of *Let-7*a microRNA by real-time PCR in gastric carcinoma. *World J. Gastroenterol*. 2007;13:2883-2888.

Zhang, L., Huang, J., Yang, N., Greshock, J., Megraw, M. S., Giannakakis, A., Liang, S., Naylor, T. L., Barchetti, A., Ward, M. R., Yao, G., Medina, A., O'brien-Jenkins, A., Katsaros, D., Hatzigeorgiou, A., Gimotty, P. A., Weber, B. L., Coukos, G. microRNAs exhibit high frequency genomic alterations in human cancer. *Proc. Natl. Acad. Sci.* U S A. 2006; 103:9136-9141.

Zhong, M., Ma, X., Sun, C., Chen, L. MicroRNAs reduce tumor growth and contribute to enhance cytotoxicity induced by gefitinib in non-small cell lung cancer. *Chem. Biol. Interact*. 2010;184:431-438.

In: MicroRNA *Let-7*
Editor: Neetu Dahiya

ISBN: 978-1-62081-152-8

Chapter 4

ROLE OF *LET-7* NETWORKS IN STEM CELL MAINTENANCE AND DIFFERENTIATION

Candida Vaz and Vivek Tanavde[1,*]
[1]Bioinformatics Institute (BII), Agency of Science Technology and Research (A*STAR), Matrix, Singapore

1. ABSTRACT

Let-7 was identified in a genetic screen in *C. elegans*. *Let-7* regulates a number of fundamental activities in the cells such as cell division, proliferation and differentiation. *Let-7* is involved in a regulatory feed back loop with LIN28, which is one of the factors required for pluripotency of cells suggesting role of *Let-7* in stem cell differentiation and maintenance. In addition to the *Let-7*/LIN28 feedback loop which is important in cancer stem cell, other *Let-7* targets such as RAS and HMGA2 are essential for self renewal of cancer stem cells and maintenance of the undifferentiated state. *Let-7* promotes differentiation by suppressing self renewal pathways in the cells. The ability of *Let-7* members to regulate multiple pathways plays an important in fine tuning of cellular processes of self-renewal and differentiation.

2. INTRODUCTION TO *LET-7* FAMILIES AND NETWORKS

1) *Let-7* Family Is the First Identified Mirna Family

Let-7 was initially identified as a heterochronic gene by forward genetics in *Caenorhabditis elegans* (Reinhart et al., 2000), after lin-4. Although it was the second to be discovered, its high functional conservation from worms to humans has provided the guidelines for the discovery of many other miRNAs and also has laid the foundations of miRNA based regulation of gene expression (Pasquinelli et al., 2000). The *Let-7* family

[*] Correspondig author: Vivek Tanavde, Bioinformatics Institute (BII), Agency of Science Technology and Research (A*STAR) 30 Biopolis Street, #07-01 Matrix, Singapore 138671, Email: vivek@bii.a-star.edu.sg.

comprises of other miRNAs that share similar seed sequences as the *Let-7*. In *C. elgans* the *Let-7* family comprises of 9 members (*Let-7*, miR-48, miR-84, miR-241, miR-265, miR-793, miR-794, miR-795 and miR-1821). In humans, the *Let-7* family comprises of 13 members (*Let-7*a-1, *Let-7*a-2, *Let-7*a-3, *Let-7*b, *Let-7*c, *Let-7*d, *Let-7*e, *Let-7*f-1, *Let-7*f-2, *Let-7*g, *Let-7*i, miR-98 and miR-202) located on nine different chromosomes.

In *C. elegans Let-7* is known to regulate the larval to adult transition. The hypodermal skin cells known as the seam cells undergo asymmetrical divisions at each larval stage. The seam cells divide in a stem cell like manner, with one daughter cell self-renewing and the other daughter cell differentiating. At the larval 4 to the adult transition, these self-renewing daughter cells stop dividing, terminally differentiate, and secrete alae. In case of loss-of-function *Let-7* mutants, these seam cells continue to divide, fail to exit the cell cycle and fail to differentiate, causing no alae formation. This leads to bursting of the vulva leading to an ultimate death of the worm. Therefore the *Let-7* gene derives its name (lethal-7) from the lethality of its absence or mutation. It has a major role in larval 4 to the adult stage transition, and is detected at the larval 3 stage reaching its maximum level at the larval 4 stage. Expression of the *Let-7* at the larval stage 2, could cause skipping of the cell-cycle leading to differentiation and premature adult development at the larval 4 stage (Abbott et al., 2005).

The other members of the *Let-7* family such as miR-48, miR-84 and miR-241 also participate in regulation of the temporal patterning at the larval 2 to larval 3 stage transition, and serve as heterechronic genes (Caygill et al., 2008). This was demonstrated by the single, double and triple mutants for these miRNAs. The single mutants displayed weak defects of extra moulting at the adult stage, which was enhanced in miR-48/miR-84, miR-48/miR-241 double mutants. The double mutants showed incomplete alae formation and lethality. These defects occurred with higher penetrance in miR-48/miR-84/miR-241 triple mutants, clearly indicating functional cooperation among them (Abbott et al., 2005). The targets of these miRNAs are additional members of the heterochronic pathway, some of which are conserved in other organisms. The Ras GTPase-family member let-60 is one such example, which is also a target for the human *Let-7*.

In *Drosophila,* the *Let-7* family has just one member, which is the *Let-7* itself. The mature sequence of the *Let-7* is 100% identical to that of the *C. elegans.* Here too the *Let-7* functions as a heterochronic gene (Caygill et al., 2008; Sikol et al., 2008), being expressed at the end of the third larval stage and peaking at the pupal stage (Pasquinelli et al., 2000). The expression of *Let-7* coincides with the release of ecdysone and is responsible for the regulation of the neuromuscular junctions in the abdominal muscles. The *Let-7* mutants display defects in maturation of the neuromuscular junctions in the abdominal muscles and exhibit juvenile features in their neuromusculature, affecting their motility, flight and fertility (Sokol et al., 2008). The *Let-7* exerts this function by down-regulation of the abrupt gene.

In *Homo sapiens*, the *Let-7* family comprises of 10 mature miRNAs produced from 13 precursor sequences. The *Let-7* family is present in multiple locations in the genome and also has several isoforms. To distinguish between the isoforms, a letter is placed after the *Let-7* to indicate a *Let-7* with a slightly different sequence, and a number at the end denotes the same sequence present at different genomic locations. The mature *Let-7*a is produced from three separate precursors, namely: *Let-7*a-1, *Let-7*a-2, *Let-7*a-3 and the mature *Let-7*f is produced from two precursors at two different genomic locations, namely *Let-7*f-1, *Let-7*f-2 (Roush and Slack, 2008).

The level of *Let-7* rises during embryogenesis (Schulman et al., 2005). Moreover, reduced expression of *Let-7* family has been associated with cancers (Park et al., 2007) leading to a conclusion that like in *C. elegans*, the *Let-7* family is involved in promoting differentiation and function as mainly tumor suppressors in humans. *Let-7* expression is undetectable in human and mouse embryonic stem cells and its level rises upon differentiation (Thomson et al., 2006).

2) *Let-7* Members Act in Concert as a miRNA Network

The *Let-7* family in humans comprising of 13 members appear to be acting together as a network, as in many cancer types the expression of various *Let-7* genes is down-regulated. The first direct piece of evidence for a role of *Let-7* in cancer was the observation that most or all of the *Let-7* family members were reduced in a significant number of lung cancer cell lines and primary human lung cancer tissues (Takamizawa et al., 2004). It was shown for the first time that *Let-7* expression is frequently reduced in lung cancers and that alterations in their expression could have a prognostic impact on the survival of surgically treated lung cancer patients.

Using miRNA microarrays, several miRNAs aberrantly expressed in human ovarian cancer tissues and cell lines were studied, among which several members of the *Let-7* family were found regulated (Dahiya et al., 2008). It was shown that the absence of *Let-7* family could be used to determine tumor origin and proliferation state because tumor suppressor miRNAs were significantly down-regulated in primary effusion lymphoma (PEL) and in Kaposi sarcoma (KS), an endothelial cell tumor.

The production of the *Let-7* family of miRNAs is known to be negatively regulated by LIN28. The embryonic cells contain LIN28 that acts as a Drosha inhibitor, preventing the transcription of the *Let-7* family members. It was observed by a study carried out in the mouse embryonal carcinoma cell line P19 that in-spite of the levels of the primary transcripts being high, the levels of the processed precursor and mature forms were low. This was because of the binding of a Drosha inhibitor to a conserved region in the loop causing inhibition of the *Let-7* processing. Differentiation of P19 with retinoic acid leads to an induction in *Let-7* caused by decay in LIN28. The similar happens during the embryonic development, when Drosha freed of LIN28 inhibition processes *Let-7* efficiently leading to an increase in *Let-7* mature species (Newman and Thomson, 2008)

Therefore, it is clear that restoration of *Let-7* processing block is essential for reprogramming stem cells, whereas a combination of LIN28 with NANOG, SOX2 and OCT3/4 can convert a human somatic cell to a pluripotent stem cell. The major regulatory controls in *Let-7* production are at the Drosha step involving LIN28, at the Dicer processing step and even at the primary transcript processing step (Wulczyn et al., 2007; Chang et al., 2008).

Several important oncogenes have been identified as the targets of *Let-7*, including RAS, MYC and HMGA2 (Chang et al., 2008; Johnson et al., 2005; Mayr et al., 2007). These target genes, making use of the RAS–MEK pathway contribute to Epithelial to Mesenchymal transition (EMT), though *Let-7* miRNA-mediated HMGA2 down-regulation had no effect on the prevention of the transformed phenotype in pancreatic cancer cells (Watanabe et al., 2008). Another *Let-7* miRNA-mediated tumor metastasis-regulating pathway was addressed recently. Raf kinase inhibitory protein, which inhibits mitogen-activated protein kinase signaling cascades, can decrease transcription of LIN28 by MYC. Suppression of LIN28 enabled it to enhance *Let-7* processing in breast cancer cells, allowing the elevated *Let-7* expression to inhibit HMGA2 that activated pro-invasive and pro-metastatic genes. As *Let-7* targets RAS, the upstream activator of RAF1, a positive feedback loop emerged to control tumor invasion and metastasis (Dangi-Garimella et al., 2009). However, there are also cases, where only some specific members of the family members are down-regulated in cancers. Although a frequent down-regulation of the *Let-7* family members is observed, there are few studies showing upregulation of some *Let-7* family members in cancer (Lawrie et al., 2009).

Therefore, in-spite of the members having a vast overlapping set of targets its not very clear, as to whether all the members of the *Let-7* family have the same function or whether they have different functions. Moreover, their regulation is not just tissue specific but also cell specific. Members of the *Let-7* family can have differential regulation within the same cell (Guled et al., 2009). This may enable *Let-7* to have differential functions in stem cells and differentiated cells. It is unclear how this differential regulation mechanism works but the different isoforms of *Let-7* primary miRNAs, the coding region and the timing of expression of these miRNAs could have a role in this regulation.

3. Cellular Functions Regulated by *Let-7* miRNA Networks

1) Role of *Let-7* in Cancer as a Tumor Suppressor

Let-7 is known to be targeting several oncogenes. The main target of *Let-7* is RAS oncogene that is found to be highly deregulated in human cancers. The 3'UTR of RAS contains several *Let-7* complementary sites (LCS) and it was found to have an inverse correlation to the *Let-7* family. In lung cancers the *Let-7* family members were found to be low as compared to RAS, which was found to be up-regulated. The role of *Let-7* in inhibiting RAS was also proved in non small cell lung cancer (NSCLC), where it was shown in a mouse model that *Let-7*g inhibited tumor growth by suppression of RAS (Kumar et al., 2008). The second major target is HMGA2 which is a chromatin-associated non-histone and a high mobility group protein having an oncogenic role in a variety of tumors including mesenchymal tumors and lung cancers. It is widely expressed in undifferentiated embryonal tissues but undetectable in adult normal tissues. It is also expressed in both benign and malignant tumors.

Ectopic expression of *Let-7* reduced HMGA2 levels and cell proliferation in lung cancer. The effect of *Let-7* on HMGA2 is dependent on multiple target sites present in its 3'UTR. It was also shown that the separation of the 3'UTR from the ORF by chromosomal translocations released the oncogene from repression. It was revealed by the study that there exists a reciprocal relation between the *Let-7* growth suppressor and the HMGA2 oncogene. In differentiated tissue there occurs an induction in *Let-7* with corresponding absence of HMGA2, similarly in lung cancers HMGA2 over expression occurs owing to reduction in *Let-7* expression or loss of the *Let-7* target sites (Mayr et al., 2007; Hebert et al., 2007; Wang et al., 2007). Around twelve oncogenes were identified to be *Let-7* regulated (Boyerinas et al.,

2010). These genes were picked out by overlap between computational prediction and experimental validation of being down-regulated when *Let-7* was up-regulated.

2) *Let-7* as a Master Regulator of Cell Proliferation

The above mentioned studies suggest that *Let-7* may control a variety of processes both during the development and maintenance of adult tissue homeostasis. In order to provide evidence for the crucial role of *Let-7* on cellular growth and proliferation in mammalian cells, the levels of *Let-7* were manipulated by using exogenously transfected pre-*Let-7* RNAs (to overexpress *Let-7*) and anti-*Let-7* oligonucleotides (to reduce the *Let-7* activity). A human lung cancer cell line (A549) and a human liver cancer cell line (HepG2) were first transfected by the synthetic miRNAs and monitored. A significant amount of proliferating cells were reduced in both the cell lines. In contrast to this, transfection of these cell lines with the antisense oligonucleotides against *Let-7* brought about a two-fold increase in proliferation. Here again, the inverse correlation between *Let-7* levels and cellular proliferation justifies the function of *Let-7* as regulator of cell division and cell survival (Johnson et al., 2007).

3) Gene Ontology (GO) Classes and Functions of the Targets of *Let-7*

In order to determine the cellular pathways regulated by *Let-7*, a microarray analysis of the A549 and HepG2 cell lines treated with *Let-7* miRNAs was carried out (Johnson et al., 2007). The differentially expressed genes were identified and were grouped by their associated biological functions using the Gene Ontology database. The primary GO classes associated with the differentially expressed genes were found to be linked to cell cycle. The GO categories were DNA replication, M phase of mitotic cell-cycle, mitotic cell cycle, cell cycle checkpoint, cell division, DNA replication initiation, mitotic check point and spindle organization and biogenesis.

It was thus concluded that the over-expression of *Let-7* causes human cancer cells to decrease cell cycle progression. *Let-7* is known to control several cell proliferation genes, which strongly suggests that *Let-7* is a key regulator of cell cycle progression. It was shown that *Let-7* directly regulates a few cell cycle proto-oncogenes like: RAS, CDC25A, CDK6 and cyclin D thus controlling cell proliferation by reducing the flux through the pathways promoting the G1 to S transition. Many of these genes are oncogenes and in cancer cells with poor levels of *Let-7*, these genes may get up-regulated which is likely to stimulate cell cycle, DNA synthesis and hence cell division.

4) Regulation of Dicer by *Let-7* through a Negative Feed Back Loop

Among the several genes known to be targets of *Let-7* family and that show an inverse correlation with its expression is the Dicer. A luciferase assay using a reporter in the 3'UTR of Dicer revealed that *Let-7* directly affects Dicer. Over expression of *Let-7* reduces the expression of Dicer, which eventually causes the reduction in the levels of other mature miRNAs along with own mature miRNA expression. Thus its involvement in a negative feed back loop helps in maintaining the equilibrium state of Dicer and various miRNAs (Forman et al., 2008; Jakymiw et al., 2010; Tokumaru et al., 2008).

4. Role of *Let-7* in Maintenance of Stemness

As seen in the earlier sections, *Let-7* regulates a number of genes involved in cell division and proliferation. We now turn to another very important aspect of *Let-7* regulation in stem cells and its impact on stem cell differentiation. Stem cell research is a cutting edge research that has given an impetus to regenerative medicine and several disease treatments, and promises to revolutionize 21st century medicine. It is therefore, crucial to understand the biology of stem cells, the network of factors required to induce and maintain stemness as well as the mechanisms that govern differentiation into specific tissues.

Stem cells express a set of factors that are responsible of their “stemness”. They have a unique transcriptional profile that has to be maintained, and which on differentiation shifts to an alternate profile. Both transcriptional regulation and epigenetic regulation play pivotal roles in maintaining the existing profile as well its plasticity and were identified by comparisons of the expression profiles of the embryonic stem cells and their differentiated derivatives (Ramalho-Santos et al., 2002).

1) Regulation by Transcription Factors

The transcription factors OCT4, NANOG, SOX2, were identified by chromatin immunoprecipitation (CHIP) coupled with microarrays and were found to regulate several genes involved in differentiation and development and maintain them at low levels in the ES cells. These three transcription factors have common set of targets and also regulate transcription of each other. In addition to OCT4, SOX2 and NANOG, many other factors required for pluripotency have been identified, including LIN28, SAL14, DAX1, ESSRB, TBX3, TCL1, RIF1, NAC1 and ZFP281. These pluripotency factors regulate each other form a complicated transcriptional regulatory network in embryonic stem (ES) cells (Wang et al., 2006).

2) Regulation by Epigenetic Factors

Besides, transcription factors, a number of epigenetic factors are also essential for maintaining the pluripotency of the ES and the somatic stem cells. As the substrate of transcription, chromatin is subjected to various forms of epigenetic regulation like chromatin remodeling, histone modifications, histone variants and DNA methylation. For example, trimethylation of lysine 9 and lysine 27 of histone 3 (H3K9 and H3K27) correlate within active regions of chromatin, whereas H3K4 trimethylation and acetylation of H3 and H4 are associated with active transcription (Jenuwein and Allis, 2001), and DNA methylation generally represses gene expression (Santos and Dean, 2004). For maintaining pluripotency, those genes whose up-regulation leads to differentiation should be inactive. Poly-comb group proteins (PcG) are functional in silencing these developmental regulators. The pluripotency factors such as OCT4, NANOG and SOX2 form a significant fraction of PcG. The genes regulated by the PcG proteins are co-occupied by nucleosomes with trimethylated H3K27. These genes are downregulated in the ES and are activated when differentiation is induced.

3) Crosstalk between Transcriptional and Epigenetic Regulation

Both the transcriptional and epigenetic pathways cross-talk with one another to maintain pluripotency. Pluripotency factors regulate genes encoding epigenetic control factors. It has been shown that OCT4, SOX2 and NANOG co-regulate certain genes encoding components of chromatin remodeling and histone modifying complexes, such as SMARCAD1, MYS3 and SET (Boyer et al., 2005). Moreover, pluripotency factors also interact with histone modifying enzymes and chromatin remodeling complexes (Wang et al., 2006). Finally, the genes of pluripotency factors are themselves subjected to epigenetic regulation (Loh et al., 2007).

Thus the interplay of the two help in activating the genes required for pluripotency and suppresses those genes that bring about development and differentiation (Chen and Daley, 2008). But in-spite of the genes controlling differentiation being transcriptionally inactive, they are maintained in a potent state for transcriptional activation. Also, the hyperdynamic chromatin structure in ES cells contributes to a rapidly plastic transcriptional profile that underlines the multi-lineage differentiation potential of ES cells. Upon differentiation, a diminution of pluripotency factors occurs that leads to dramatic changes in the transcriptional profile, where just the reverse begins to occur; the genes controlling differentiation get active, the pluripotency causing genes become suppressed and the chromatin structure gets compact. Post transcriptional regulation of factors in both transcriptional and epigenetic regulatory networks adds another layer of regulation towards this process. Such complex layers of regulation are necessary to achieve the precise temporal and spatial gene expression necessary for the stem cells to differentiate. MiRNAs are important post transcriptional regulators and could play a role in regulating the transcriptional and epigenetic networks in stem cells.

4) Evidence of the Role of *Let-7* in Regulation of Pluripotency

Additionally, microRNAs have emerged as key regulators of pluripotency (Gunaratne, 2009; Mallanna et al., 2010). Yu et al., (2007) demonstrated that self renewal and maintenance of the undifferentiated state required reduced *Let-7* miRNA levels. They showed this in cancer stem cells isolated from breast tumor xenografts. This requirement appears to be mediated through HMGA2 and RAS proteins. RAS targeting by *Let-7* miRNAs appeared to be essential for self renewal of cancer stem cells whereas HMGA2 regulation seemed to be important for maintenance of the undifferentiated state. This mechanism of HMGA2 mediated regulation of stemness maintenance in cancer stem cells appears to be similar to *Let-7*/HMGA2 mechanism in embryonic stem cells. Although Yu et al. (2010) have shown that antisense targeting of *Let-7* in cancer stem cells triggered de-differentiation of a breast tumor cell line, this study provides no direct evidence that lack of *Let-7* miRNAs is essential for maintenance of stemness. However the data from this study provide very strong circumstantial evidence in favor of this hypothesis.

More recently Melton et al (2010) demonstrated that *Let-7* overexpression led to differentiation of embryonic stem cells in the absence of DGCR8 (a gene essential for miRNA maturation). However this effect was not seen in the wild type embryonic stem cells with normal levels of DGCR8. The introduction of the ES cell cycle regulating (ESCC) miRNAs into DGCR8 -/- cells rescued the cell cycle defects and promoted the self-renewal tendency. On the other hand, the introduction of the *Let-7* miRNAs suppressed self renewal in these cells. In wild type cells, however the ESCC miRNAs inhibit the capacity of *Let-7* to silence self-renewal.

These findings suggested that the "functional antagonism" between *Let-7* and ESCC miRNA families causes them to have opposing effects on ESC self-renewal. In order to find out the mechanism for the opposing roles; a pathway analysis on the miRNA regulated transcripts was carried out. Overlaps between the miRNA regulated genes and CHIP detected pluripotency transcription factors. Around 88 transcription factors down-regulated by *Let-7* were found to be up-regulated by ESCC miRNAs. These transcription factors included pluripotency genes such as LIN28, SALl4, n-MYC and c-MYC. qRT-PCR, western analysis and reporter assays also confirmed the opposing effects of *Let-7* and ESCC on the levels of LIN28, SALl4, n-MYC and c-MYC.

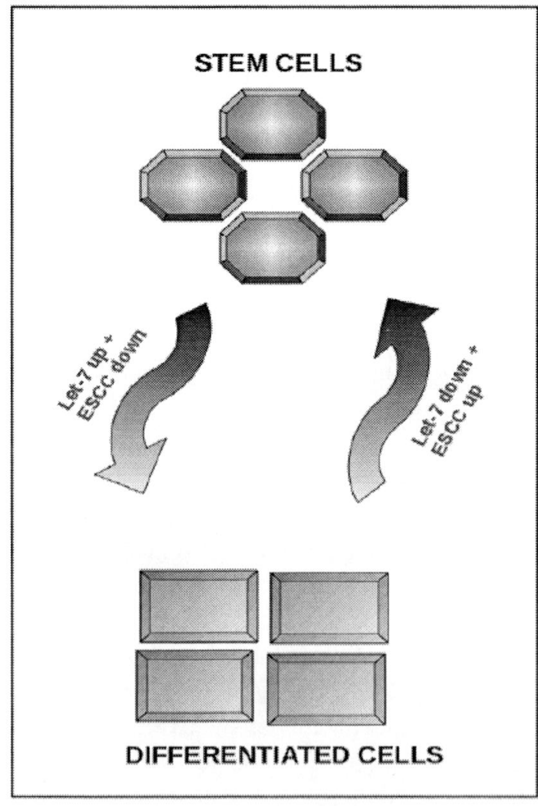

Figure 1. *Let-7* levels determine stemness and differentiation status of embryonic stem cells. *Let-7* and ESCC miRNAs act in an antagonistic manner to trigger differentiation or maintain stemness of ES cells.

Let-7 promotes differentiation by suppressing the self-renewal pathway, whereas the ESCC miRNAs promote the self-renewal pathway. Inhibition of *Let-7* promotes de-differentiation and can reprogram the somatic cells to induced pluripotent cells (iPS cells) (Figure 1). This is a good example of two different miRNA networks regulating the same gene network in an antagonistic manner.

Both of them act in self-reinforcing loops to maintain the self renewing versus the differentiated states. In the self-renewing state, the ESCC miRNAs increase the expression of LIN28 and c-MYC. LIN28 blocks *Let-7* expression and lowers it levels. The upregulated c-MYC along with the transcription factors OCT4, SOX2, NANOG activate the ESCC miRNAs further and promote the undifferentiated state. As the ES cells differentiate, OCT4, SOX2, NANOG are downregulated, resulting in loss of the ESCC miRNAs and LIN28. With the loss of LIN28, levels of *Let-7* rise rapidly. The increase in *Let-7* levels brings about down-regulation of its own inhibitor LIN28 as well as of MYC. Furthermore *Let-7* inhibits the downstream targets of OCT4, NANOG, SOX2, TCF3 to maintain the differentiated state.

5. ROLE OF *LET-7* IN ADULT STEM CELLS

Evidence for the role of *Let-7* family in maintenance of differentiation state also comes from the role of this miRNA family in adult stem/progenitor cells. *Let-7* family miRNAs seem to influence differentiation of adult stem cells through many RNA binding proteins. Kawahara et al. (2011) showed that Musashi1 in co-operation with LIN28 blocks the biogenesis of the entire *Let-7* miRNA family in Neuronal Stem/Progenitor cells. This regulation of *Let-7* biogenesis was found to be an important event in the differentiation of neuronal precursors from embryonic stem cells. Further this suppression of *Let-7* miRNA levels may also contribute to maintain the undifferentiated state of the neuronal stem/progenitor cells.

The LIN28/*Let-7* feedback loop has also shown to be important in cancer stem cells expression of high levels of Aldehyde Dehydrogenase (ALDH) (an enzyme that has been used as a functional stem cell marker in the hematopoietic system) (Yang et al., 2010). In this study inhibition of *Let-7* expression by LIN28 was found to be a pre-requisite for maintenance of stemness in ALDH+ cancer stem cells. Overexpression of *Let-7* led to decreased numbers of ALDH+ cancer stem cells. However it was unclear from this study if the cancer stem cells differentiated to a more mature phenotype. Since differentiation is one of the key end points of current cancer therapy, the *Let-7*/LIN28 mechanism could be used as a target for exhausting cancer stem cells within a tumor. This study also demonstrates that the same mechanism is active in differentiation of normal mammary gland progenitor cells.

Suppression of *Let-7* activity also leads to proliferation of progenitors and stem cells in the hematopoietic system. Recently Ikeda et al. (2011) have shown that deletion of HMGA2 binding site of *Let-7* miRNAs leads to increase in erythroid and myeloid progenitors as well as KSL stem cells. Absence of *Let-7* miRNA binding site also led to erythroid progenitors which were growth factor independent compared to wild type progenitors that needed erythropoietin for their survival. Thus absence of *Let-7* seems to favor a proliferative phase in these cells.

Our group has shown that unlike most other stem/progenitor cells, *Let-7* family members are highly expressed in Mesenchymal Stromal Cells (MSCs) (Koh et al., 2010). It is unclear why *Let-7* is highly expressed in MSCs when it is suppressed in all other stem cell types. This might reflect the more differentiated nature of MSCs (more progenitor and not a stem cell phenotype) or it is possible that *Let-7* acts via different mechanism to promote MSC self renewal.

The *Let-7* family miRNAs do not seem to play a role in osteochondrogenic or adipogenic differentiation of MSC. However, our study shows the ability of the *Let-7* miRNA network to regulate a gene network possibly involved in hepatic differentiation of MSC. The transcription factor HNF4A is at the hub of this network (Figure 2). Interestingly HNF4A is not predicted to be directly regulated by any miRNA, but the ability of the *Let-7* miRNA network to indirectly regulate the expression of HNF4A further supports the idea that *Let-7* miRNA network can act in concert to regulate cellular differentiation.

It is interesting to note that most genes in this network are suppressed when *Let-7* family miRNAs are over-expressed (MSC) and up-regulated when *Let-7* family gene expression goes down (HEPG2 cells).

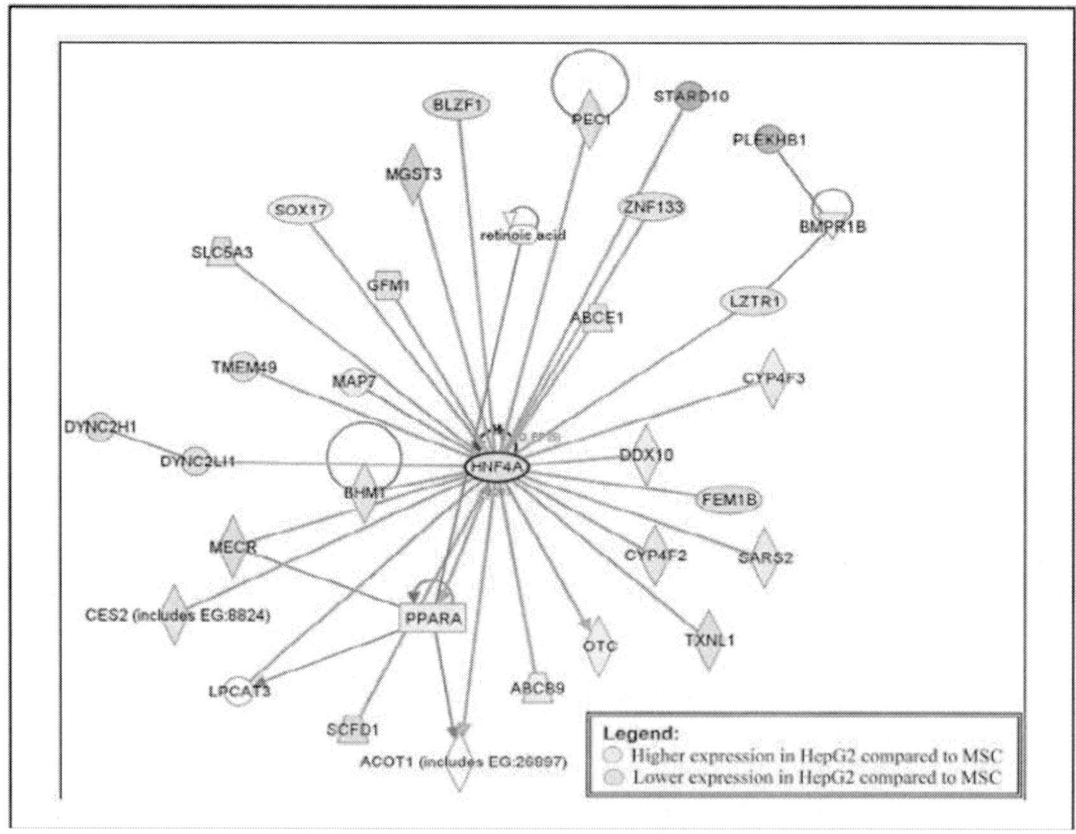

Figure 2. *Let-7* family miRNAs regulate a gene network with HNF4A as its hub.

CONCLUSION

Thus there is increasing evidence that the *Let-7* family of miRNAs regulate multiple gene networks critical in maintenance of the stem cell phenotype as well as differentiation of stem cells to desired lineages. It is unlikely that *in vivo* miRNAs act as single entities. Increasingly gene networks and pathways have been shown to be regulated by miRNAs belonging to a single family. The *Let-7* family members regulate critical genes involved in cellular proliferation and differentiation. This unique ability to regulate multiple gene networks simultaneously gives the *Let-7* family miRNAs the ability to regulate the cellular processes of self renewal and differentiation. It will be interesting to study whether *Let-7* family members use similar mechanisms for such regulation especially in different adult stem/progenitor cells and embryonic stem cells. Also *Let-7* family members might interact with other miRNA networks to achieve this regulation of stem cell maintenance. Understanding how these networks interact will provide new insights into miRNA regulation of stem cell differentiation.

It will be interesting to explore the role of *Let-7* miRNAs in self renewal and differentiation of stem/progenitor cells from different tissues. Although in embryonic stem cells levels of *Let-7* correlate inversely with the differentiation status of the cell, our data from MSC shows this may not be the case in other stem cells; especially adult stem cells. The role of *Let-7* family miRNAs in maintenance of stem cells from intestines, skin, muscle and other tissues remains to be explored. Exploring the role of *Let-7* regulation in these cells may lead to novel mechanisms of *Let-7* mediated regulation that are different from the HMGA2 or LIN28 mediated mechanisms.

REFERENCES

Abbott, A. L. et al. The *Let-7* MicroRNA family members miR-48, miR-84, and miR-241 function together to regulate developmental timing in *Caenorhabditis elegans*. *Developmental Cell* 2005;9:403-414.

Boyer, L. A. et al. Core transcriptional regulatory circuitry in human embryonic stem cells. *Cell* 2005;122:947-956.

Boyerinas, B., Park, S. M., Hau, A., Murmann, A. E., P. ME. The role of *Let-7* in cell differentiation and cancer. *Endocr. Relat. Cancer* 2010;17:F19-36.

Caygill, E. E., Johnston, L. A. Temporal regulation of metamorphic processes in Drosophila by the *Let-7* and miR-125 heterochronic microRNAs. *Current Biology* 2008;18:943-950.

Chang, T. C. et al. Widespread microRNA repression by Myc contributes to tumorigenesis. *Nature Genetics* 2008;40:43-50.

Chen, L., Daley, G. Q. Molecular basis of pluripotency. *Hum. Mol. Genet.* 2008;17:R23-27.

Dahiya, N. et al. MicroRNA expression and identification of putative miRNA targets in ovarian cancer. *PLoS One* 2008;3:e2436.

Dangi-Garimella, S. et al. Raf kinase inhibitory protein suppresses a metastasis signalling cascade involving LIN28 and *Let-7*. *EMBO Journal* 2009;28:347-358.

Forman, J. J., Legesse-Miller, A., Coller, H. A. A search for conserved sequences in coding regions reveals that the *Let-7* microRNA targets Dicer within its coding sequence. *Proc. Natl. Acad. Sci.* U S A 2008;105:14879-14884.

Guled, M. et al. CDKN2A, NF2, and JUN are dysregulated among other genes by miRNAs in malignant mesothelioma -A miRNA microarray analysis. *Genes Chromosomes Cancer* 2009;48:615-623.

Gunaratne, P. H. Embryonic stem cell microRNAs: defining factors in induced pluripotent (iPS) and cancer (CSC) stem cells? *Curr. Stem. Cell Res. Ther.* 2009;4:168-177.

Hayes, G. D., Ruvkun, G. Misexpression of the *Caenorhabditis elegans* miRNA *Let-7* is sufficient to drive developmental programs. *Cold Spring Harbor Symposia on Quantitative Biology* 2006;71:21-27.

Hebert, C., Norris, K., Scheper, M. A., Nikitakis, N., S. J. J. High mobility group A2 is a target for miRNA-98 in head and neck squamous cell carcinoma. *Molecular Cancer* 2007;6:5.

Ikeda, K., Mason, P. J., Bessler, M. 3'UTR-truncated Hmga2 cDNA causes MPN-like hematopoiesis by conferring a clonal growth advantage at the level of HSC in mice. *Blood* 2011;117:5860-5869.

Jakymiw, A. et al. Overexpression of dicer as a result of reduced *Let-7* MicroRNA levels contributes to increased cell proliferation of oral cancer cells. *Genes Chromosomes Cancer* 2010;49:549-559.

Jenuwein, T., Allis, C. D. Translating the histone code. *Science* 2001;293:1074-1080.

Johnson, C. D. et al. The *Let-7* microRNA represses cell proliferation pathways in human cells. *Cancer Research* 2007;67:7713-7722.

Johnson, S. M et al. RAS is regulated by the *Let-7* microRNA family. *Cell* 2005;120:635-647.

Kawahara, H. et al. Musashi1 Cooperates in Abnormal Cell Lineage Protein 28 (LIN28)-mediated *Let-7* Family MicroRNA Biogenesis in Early Neural Differentiation. *J. Biol. Chem.* 2011;286:16121-16130.

Koh, W. et al. Analysis of deep sequencing microRNA expression profile from human embryonic stem cells derived mesenchymal stem cells reveals possible role of *Let-7* microRNA family in downstream targeting of hepatic nuclear factor 4 alpha. *BMC Genomics* 2010;11:S6.

Kumar, M. S. et al. Suppression of non-small cell lung tumor development by the *Let-7* microRNA family. *Proc. Natl. Acad. Sci.* U S A 2008;105:3903-3908.

Lawrie, C. H. et al. Expression of microRNAs in diffuse large B cell lymphoma is associated with immunophenotype, survival and transformation from follicular lymphoma. *Journal of Cellular and Molecular Medicine* 2009;13:1248-1260.

Loh, Y. H., Zhang, W., Chen, X., George, J., Ng, H. H. Jmjd1a and Jmjd2c histone H3 Lys 9 demethylases regulate self-renewal in embryonic stem cells. *J. Genes Dev.* 2007;21:2545-2557.

Mallanna, S. K., Rizzino, A. Emerging roles of microRNAs in the control of embryonic stem cells and the generation of induced pluripotent stem cells. *Dev. Biol.* 2010;344:16-25.

Mayr, C., Hemann, M. T., Bartel, D. P., Disrupting the pairing between *Let-7* and Hmga2 enhances oncogenic transformation. *Science* 2007;315:1576-1579.

Melton, C., Judson, R. L., Blelloch, R. Opposing microRNA families regulate self-renewal in mouse embryonic stem cells. *Nature* 2010;463:621-626.

Newman, M. A., Thomson, J. M., H. SM. LIN28 interaction with the *Let-7* precursor loop mediates regulated microRNA processing. *RNA* 2008;14: 1539-1549.

Park, S. M. et al. *Let-7* prevents early cancer progression by suppressing expression of the embryonic gene HMGA2. *Cell Cycle* 2007;6:2585-2590.

Pasquinelli, A. E. et al. Conservation of the sequence and temporal expression of *Let-7* heterochronic regulatory RNA. *Nature* 2000;408:86-89.

Ramalho-Santos, M., Yoon, S., Matsuzaki, Y., Mulligan, R. C., Melton, D. A. "Stemness": transcriptional profiling of embryonic and adult stem cells. *Science* 2002;298:597-600.

Reinhart, B. J. et al. The 21-nucleotide *Let-7* RNA regulates developmental timing in Caenorhabditis elegans. *Nature* 2000;403:901-906.

Roush, S., Slack, F. J. The *Let-7* family of microRNAs. *Trends in Cell Biology* 2008;18:505-516.

Santos, F., Dean, W. Epigenetic reprogramming during early development in mammals. *Reproduction* 2004;127:643-651.

Schulman, B. R., Esquela-Kerscher, A., Slack, F. J. Reciprocal expression of lin-41 and the microRNAs *Let-7* and miR-125 during mouse embryogenesis. *Developmental Dynamics* 2005;234:1046-1054.

Sokol, N. S., Xu, P., Jan, Y. N., A. V. Drosophila *Let-7* microRNA is required for remodeling of the neuromusculature during metamorphosis. *Genes and Development* 2008;22:1591-1596.

Takamizawa, J. et al. Reduced expression of the *Let-7* microRNAs in human lung cancers in association with shortened postoperative survival. *Cancer Research* 2004;64:3753-3756.

Thomson, J. M. et al. Extensive post-transcriptional regulation of microRNAs and its implications for cancer. *Genes Development* 2006;20:2202-2207.

Tokumaru, S., Suzuki, M., Yamada, H., Nagino, M., Takahashi, T. *Let-7* regulates Dicer expression and constitutes a negative feedback loop. *Carcinogenesis* 2008;29:2073-2077.

Wang, J. et al. A protein interaction network for pluripotency of embryonic stem cells. *Nature* 2006;444:364-368.

Wang, T. et al. A micro-RNA signature associated with race, tumor size, and target gene activity in human uterine leiomyomas. *Genes Chromosomes Cancer.* 2007;46:336-347.

Watanabe, S. et al. HMGA2 maintains oncogenic RAS-induced epithelial–mesenchymal transition in human pancreatic cancer cells. *The American Journal of Pathology* 2009;174:854-868.

Wulczyn, F. G. et al. Post-transcriptional regulation of the *Let-7* microRNA during neural cell specification. *FASEB J.* 2007;21:415-426.

Yang, X. et al. Double-negative feedback loop between reprogramming factor LIN28 and microRNA *Let-7* regulates aldehyde dehydrogenase 1-positive cancer stem cells. *Cancer Res.* 2010;70:9463-9472.

Yu, F. et al. *Let-7* regulates self renewal and tumorigenicity of breast cancer cells. *Cell* 2007;131: 1109-1123.

In: MicroRNA *Let-7*
Editor: Neetu Dahiya

ISBN: 978-1-62081-152-8

Chapter 5

THE ROLE OF MICRORNA *LET-7* IN STEM/PROGENITOR CELLS

***Shao-Chih Chiu*[1,2,3,*] *and Shinn-Zong Lin*[1,2,3]**

[1]Graduate Institute of Immunology, China Medical University, Taichung, Taiwan

[2]Center for Neuropsychiatry, China Medical University Hospital, Taichung, Taiwan

[3]Department of Neurosurgery, China Medical University, Beigan Hospital, Yunlin, Taiwan

1. ABSTRACT

One of the first discovered human microRNAs, *Let-7* and its family members, are highly conserved across species in the regulation of embryonic development and stemness. Misregulation of *Let-7* results in a dedifferentiated cellular state and the development of cell-based disorders such as cancer. Although, many preclinical studies have been devoted to the *Let-7*-based cancer therapy, research into what is the physiological function of *Let-7* in the regeneration of adult tissues has only just begun. The exact role of *Let-7* in cancer is not yet fully understood. There is a need to understand the biological meanings of *Let-7* alterations in normal stem cells and cancers before proceeding to clinical applications. This chapter highlights an outlook on certain pivotal aspects of *Let-7* in stemness and tumorigenesis for future therapeutic potential.

2. INTRODUCTION

MicroRNAs (miRNAs) are short non-coding and small endogenous (~22 nucleotide) RNAs that typically imperfectly base pair with 3' untranslated regions (3'UTRs) and mediate translational repression and mRNA degradation (for reviews, see Refs (Nilsen, 2007; Stefani

* Corresponding author: Shao-Chih Chiu, Graduate Institute of Immunology, China Medical University, 91, Hsueh-Shih Rd.,TaiChung, Taiwan, 40402 Email: scchiu@mail.cmu.edu.tw.

and Slack, 2008). One of the earliest found miRNAs, lethal-7 (*Let-7*), is shown to posses the essential role in the development of *Caenorhabditis elegans* and *Drosophila melanogaster* (Pasquinelli et al., 2000; Reinhart et al., 2000), and subsequent work has shown that both its sequence and function are highly conserved in mammals. During general development, accumulation of *Let-7* can modulate activities of pluripotent factors, such as LIN28, and determine the timing of differentiation. Based on recent findings, it is revealed that *Let-7* regulates 'stemness' by repressing self-renewal and promoting differentiation in normal development for stem cells/progenitors.

Many of state-of-the-art miRNAs studies are focused on *Let-7* due to its plentiful and pivotal roles in cell proliferation, reprogramming, development, differentiation and tumorigenesis (for reviews, see Refs (Boyerinas et al., 2008; Bussing et al., 2008; Dahiya and Morin, 2010; O'Day and Lal, 2010; Ortholan et al., 2009; Roush and Slack, 2008). The major function of *Let-7* is to regulate the stem cells/progenitors into differentiated cells and to reprogram cell fates, and is necessary to be noted and discussed. It is still vague about the naturally physiological or biological performance of *Let-7* family. Recently, increasing evidences reveal that *Let-7* acts as a tumor suppressor by targeting various oncogenes, cell-cycle regulators and cell survival pathways. Most of studies indicate that the expression of *Let-7* in various cancer stem-like cells is notable lower than it in differentiated counterpart cells. Several studies focused on how to regulate and manipulate the expression of *Let-7*, will be valuable in helping us to understand the regulation of cell fates, especially in stem cells/progenitors.

This chapter presents a viewpoint of *Let-7* in aspect of normal embryonic and adult stem cells and discusses the critical issues that should be examined for the further development of *Let-7*-based therapy.

3. The Role of *Let-7* in Normal Stem/Progenitor Cells and Cancer Cells

Mature *Let-7* family members are absent in embryonic stem (ES) cells or pluripotent progenitors, and then increasing expression of functional *Let-7* upon differentiation seems to be a common scenario (Figure 1) (Hermann et al., 2010). These properties are also shared with tumor initiating cells, and loss of *Let-7* might promote transformation of normal proliferating or dedifferentiated somatic cells into tumor initiating cells. Such loss might be caused by chromosomal abnormalities, epigenetic alterations or impaired processing of *Let-7* at the post-transcriptional level. Alternatively, normal stem cells can be transformed directly to cancer initiating cells under high self-renewing rates without normal surveillance (Figure 1). Low levels of *Let-7* are considered a notable feature of certain stem cell populations.

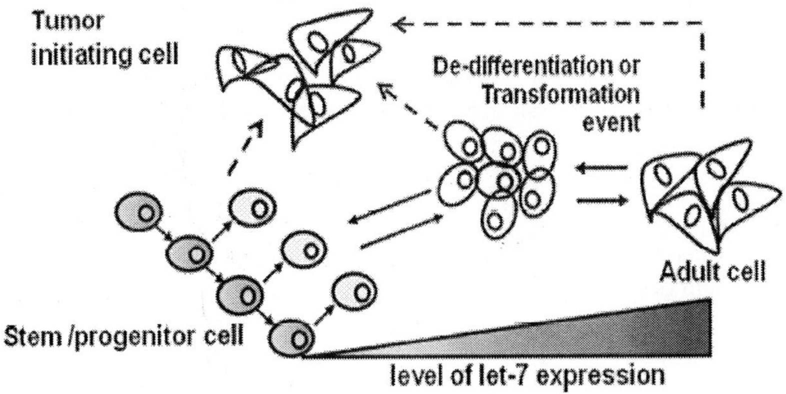

Figure 1. *Let-7* in stem cells and cancer initiating cells. During differentiation process, stem/progenitor cells become more restricted to specific cell lineages, and the increasing level of *Let-7* is shown in those differentiated cells. Broken arrows indicate hypothetical sources of cancer stem cells: loss of *Let-7* might promote transformation of normal proliferating or de-differentiated cells into cancer stem cells.

1) Current Studies of *Let-7* in the Development and Differentiation of Stem/Progenitor Cells

Stem/progenitor cells have a unique ability to replicate infinitely (self-renewal), yet they are capable of differentiating into various cell types (pluripotency). These properties of stem/progenitor cells offer immense potential for the field of regenerative medicine. However, before stem cell therapies are applied clinically, a thorough understanding of the mechanisms that control pluripotent stem cell identity would be extremely meritorious. Recent work has demonstrated that *Let-7* family members play key roles in controlling the timing of differentiation or reprograming transition of various cells via complex regulatory mechanisms or targets. Furthermore, *Let-7* family has also been shown to govern endogenous activities in various different adult cells (Table 1).

In *Drosophila*, *Let-7* determines the proper timing for cell-cycle exit, metamorphosis, neuromusculature remodeling, juvenile-to-adult-stage transition, and adult behavior (Caygill and Johnson, 2008; Sokol et al., 2008). As seen in the *Drosophila* model of Parkinson's disease, the increasing level of *Let-7* attenuates the pathogenic leucine-rich repeat kinase 2 (LRRK2) effect by targeting mRNAs of E2F1 and DP (Gehrke et al., 2010). The zebrafish ortholog of *Let-7* exhibits basal expression in the uninjured retina to suppress the expression of regeneration-associated genes and to block the premature Müller glia dedifferentiation (Ramachandran et al., 2010). In the adult newt, *Let-7* regulates the trans-differentiation and regeneration of lens and pigment epithelial cells (Nakamura et al., 2010; Tsonis et al., 2007).

More and more experimental outcomes from mammals are accumulated and examined the activity of *Let-7* in the development and physiological function (Table 2). In mice, *Let-7* has been shown to regulate the neural lineage specificity in ES cells and affect the brain development (Schwamborn et al., 2009; Sempere et al., 2004; Wulczyn et al., 2007; Zhao et al., 2010). Moreover, mammary gland epithelial progenitor cells with the non-detectable level of *Let-7* keep progenitors in the status of self-renewing and reconstitute the mammary gland by modulation of LIN28/*Let-7* loop (Ibarra et al., 2007; Yang et al., 2010). The expression of *Let-7* family triggers the formation of embryoid body and differentiation of three germ layers by multiple mechanisms in human (Ivey et al., 2008; Van Wynsberghe et al., 2011) and mice ES cells (Melton et al., 2010; Rybak et al., 2009; Zhong et al., 2010).

Table 1. *Let-7* expressed in adult cells

Cells	microRNA let-7 families	Biological activities	Targets of *Let-7*	Refs.
3T3-L1 cells, Swine adipocytes	*Let-7*a, b, d, e, g, i	To be up-regulated during adipogenesis	HMGA2	(Li et al., 2011; Sun et al., 2009)
Cardiomyocy-tes	*Let-7*, miR-98	To suppress the cardiac hypertrophy	Cyclin D2	(Yang et al., 2011)
Dopaminergic neurons	*Let-7*	To attenuate pathogenic LRRK2 effects	E2F1, DP	(Gehrke et al., 2010)
Endothelial cells	*Let-7*f	To promote angiogenesis	Thrombospond-in-1	(Kuehbacher et al., 2007)
Liver cells	*Let-7*a, b, c	To be highly expressed in adult liver development	TGF beta-R1	(Tzur et al., 2009)
Lung cells	*Let-7*a, c	To be up-regulated in allergic asthma and repress cell proliferation	IL-13, RAS	(Johnson et al., 2007; Polikepahad et al., 2010)
Mammary epithelial progenitors	*Let-7*	To differentiate progenitors, decrease colony numbers, and reduce self-renewing cells	LIN28/*Let-7* regulatory loop	(Ibarra et al., 2007; Yang et al., 2010)
Müller glia	*Let-7*a, f	To inhibit Müller glia dedifferentiation	ascl1a, hspd1, LIN28, oct4, pax6b and c-myc	(Ramachandran et al., 2010)
Ovary cells	*Let-7*d, i	To be highly expressed in newborn ovary and promote ovarian development and early folliculogenesis	N/A	(Ahn et al., 2010)
Peripheral blood	*Let-7*a, b, c, d, g, i, miR-98	To increase the sensitivity of Fas-	Fas	(Vaz et al., 2010; Wang et

Cells	microRNA let-7 families	Biological activities	Targets of *Let-7*	Refs.
mononuclear cells		mediated apoptosis		al., 2011)
Reticulocytes	*Let-7*a, c, d, e, g, i	To be up-regulated during the fetal-to-adult developmental transition in reticulocytes.	N/A	(Noh et al., 2009)
Skeletal muscle cells	*Let-7*b, e	To reduce muscle cell renewal and impair cell cycle function.	Association with CDK6, CDC25A and CDC34	(Drummond et al., 2011)
Somatic cells, *in vivo*	*Let-7*	To increase body size, crown-rump length and delayed onset of puberty	LIN28a, c-Myc, Kras, Igf2bp1 and Hmga2	(Zhu et al., 2010)
T cells	*Let-7*a, b, c, d, g	To inhibit the memory fate determination of T cells	N/A	(Almanza et al., 2010)

N/A, not applicable.

Let-7 family is also shown to play a key role in the generation and development of different adult tissue cells, including the adipogenesis in 3T3-L1 and adipocytes (Li et al., 2011; Sun et al., 2009), the cardiac hypertrophy by cardiomyocytes (Yang et al., 2011), angiogenesis of endothelial cells (Kuehbacher et al., 2007), liver development (Tzur et al., 2009), manipulation of body size and puberty (Zhu et al., 2010) and newborn ovarian development (Ahn et al., 2010). In the circulating blood system, *Let-7* is shown not only to control the developmental of reticulocytes (Noh et al., 2009) and T cells (Almanza et al., 2010) but also the increasing sensitivity of Fas-mediated apoptosis in peripheral blood mononuclear cells (Vaz et al., 2010; Wang et al., 2011).

The regulatory mechanism of *Let-7* expression is very important for the maintenance of stem/progenitor cells pluripotency as well as for reprogramming somatic cells into induced pluripotent stem cells. LIN28, which sustains the stemness state of various stem cells, is a pivotal target of *Let-7* and is downregulated by *Let-7* during the differentiation (Boyerinas et al., 2010; Bussing et al., 2008; Viswanathan et al., 2008). Recently, LIN28 and *Let-7* have been shown to have opposing expression patterns for the reciprocal regulation and functions in development and cell-fate transition. LIN28 modulates the production of endogenous pri-*Let-7* and pre-*Let-7* transcript in ES cells. It has been shown that LIN28 could block the function of either Drosha or Dicer in the processing of *Let-7* and act as a posttranscriptional repressor for the *Let-7* biogenesis (Heo et al., 2008; Viswanathan et al., 2010; West et al., 2009; Wulczyn et al., 2007).

Table 2. *Let-7* expressed in stem/progenitor cells

Cells	microRNA let-7 families	Biological activities	Targets of *Let-7*	Refs.
Embryonic stem cells	*Let-7*, b, c	To promote ES cells differentiation, EB formation	LIN28, c-Myc, n-Myc, lin-41	(Ivey et al., 2008; Melton et al., 2010; Rybak et al., 2009; Van et al., 2011; Zhong et al., 2010)
Mesenchymal stem cells	*Let-7*a, c, e, f, g, i	To suppress the expression of HNF4A	HNF4A	(Koh et al., 2010)
Neural stem/ progenitor cells	*Let-7*a, b	To inhibit proliferation of neural stem cell and promote differentiation	c-Myc, TLX, cyclin D1	(Balzer et al., 2010; Schwamborn et al., 2009; Sempere et al., 2004; Wulczyn et al., 2007; Zhao et al., 2010)
Testis-derived male germ-line stem cells	*Let-7*a, d	To be upregulated and as miRNA signatures	IGF2, H19	(Jung et al., 2010)

The recent study is shown that another key gene, HMGA2, for the self-renewal and the maintenance of stemness in adult stem cells is highly expressed in hematopoietic, adipotic and neural progenitors. During differentiation, the increasing level of *Let-7* directly targets HMGA2 to reduce the expression, following the inhibition of cell proliferation (Cavazzana-Calvo et al., 2010; Lee et al., 2007; Nishino et al., 2008; Sun et al., 2009).

Recently, some molecules associated with actions of cell cycle, cell proliferation and apoptosis are shown to be intercepted by *Let-7*. Insulin-like growth factor 2 mRNA-binding proteins (IMP1), an oncofetal gene expressed during the early fetal life, regulates the proliferation in stem cells and is targeted by *Let-7* for the blockade of MYC and K-RAS expression (Boyerinas et al., 2008; Mongroo et al., 2011). Genomic studies by the microarray analysis in cancer cells revealed that *Let-7* inhibits multiple cell-cycle-associated genes, including CDC25A, cyclin D1, D3 and A, and CDK4, 6 (Johnson et al., 2007; Schultz et al., 2008). Interestingly, the expression of *Let-7*a can suppress the doxorubicin- and paclitaxel-induced apoptosis by targeting caspase-3 in cancer cell lines, A431 and HepG2 (Tsang et al., 2008). In summary, *Let-7* family represses many key genes associated with cell cycle regulators such as CDC25A and CDK6, early expressed embryonic genes including LIN28, HMGA2, MLIN41 and IMP1 and growth and proliferation related genes including RAS and c-MYC (for reviews, see Refs (Bussing et al., 2008)). The effects of *Let-7* in ES cells can be shown its inhibitory effect on the expression of LIN28, c-MYC, SALl4 and downstream genes of pluripotency factors, in particular SOX2, OCT4, and NANOG. However, *Let-7* delivered into wild-type ES cells failed to induce ES cells to differentiate due to the antagonizing effects of other miRNAs, such as ES cell-specific cell cycle-regulating miRNAs (ESCC) in ES cells (Melton et al., 2010).

2) *Let-7* as a Tumor Suppressor in Cancer Cells

Although many studies suggest that *Let-7* regulates the proliferation, apoptosis and differentiation in cancer stem cells and can therefore be evaluated as a possible target for cancer therapies, the conflict of *Let-7* for the progression of cancer cell is observed and indeed required the further investigation. The transcriptional and post-transcriptional regulation of *Let-7* is still unclear, and *Let-7* targeting on multiple genes reveals that the specificity of *Let-7* is necessary to be addressed. In human malignant cholangiocytes, increasing levels of *Let-7*a contribute to the survival effects of enforced IL6 activity and the constitutively increased phosphorylation of STAT3 by a mechanism involving the neurofibromatosis 2 gene (Meng et al., 2007). The highly expressed *Let-7* has been shown to increase the chemoresistance of hepatocellular cancer stem cells through a novel regulatory mechanism of *Let-7*/miR-181s (Meng et al., 2011). In hepatocellular carcinoma HepG2 and squamous carcinomas A431 cells, *Let-7*a targets on caspase-3 for increasing the activity of chemo-resistance to apoptosis (Tsang et al., 2008). *Let-7*b, expressed in diffuse large B cell lymphomas, epigenetically downregulates the tumor suppressor gene, PRDM1/BLIMP1 (Nie et al., 2010). In leukemia cells, *Let-7* isoforms, -a, -b, -c, and- d, are expressed in the higher levels than in normal peripheral blood mononuclear cells (Vaz et al., 2010). *Let-7* family is selectively secreted into the extracellular environment via exosomes in a metastatic gastric cancer cell line and, thus, may be a useful marker for non-invasive screening strategies (Ohshima et al., 2010). The further investigation on the role of *Let-7* and its regulation in tumorigenesis is necessary and urgent for the miRNA-based cancer therapy.

4. The Conflict of *Let-7* between Self-Renewed and Tumor-Suppressed Activities

As accordant with the physiological function of *Let-7* family in promoting the differentiation state, miRNA-based cancer gene therapy offers the theoretical theme of targeting multiple genes and interfering with networks controlled by *Let-7*. Reconstitution of tumor-suppressive activity of *Let-7* has produced favorable antitumor outcomes in experimental models. Pending subjects need to be discussed and resolved prior to the consideration of *Let-7*-based cancer gene therapy. It contains several issues including the need for the specificity of mRNA targeted by *Let-7* in various cancers, the incomplete knowledge of regulatory mechanism and biogenesis of *Let-7* that affect the therapeutic efficiency, the possibility for nonspecific immune activation and the lack of a defined, optimal system of delivery. The limitation to tumor-suppressor *Let-7* therapy is the paucity of *Let-7*-targeted genes known to induce or preserve malignant phenotypes; moreover, more than one genetic alterations are necessary for carcinogenesis. Many differentiated tumors with average or high expression of *Let-7* are reported. Also, as it is true for many procedures of gene therapies, the eradication of treated tumors is rare even in experimental systems because of the technical difficulty of transducing sufficiently large proportions of cells in the tumors.

1) Indefinite Knowledge of *Let-7* Biology and Complicated Regulatory Mechanism

Although restoration of *Let-7* expression provides the possibility for the future cancer therapy, limited knowledge concerning its transcriptional, post-transcriptional and processing control during biogenesis of *Let-7* and tumorigenesis make it difficult to directly apply *Let-7* as a therapeutic strategy. It is necessary to confirm that the downregulation of *Let-7* in tumors is a primarily pathogenic factor during tumorigenesis. Supporting the hypothesis for cancer stem cells, many studies convey the opinion that the epigenetic downregulation of *Let-7* in cancer stem/progenitor cells is common and leads the upregulation of oncofetal genes (HMGA2, LIN28, RAS, MYC.) and, thereby, to progress of stemness activity and tumorigenesis. Notwithstanding many data have been shown the tumor-suppressive role of *Let-7*, the high expression of *Let-7* in tumors is still observed and would be beneficial to survival signals (Meng et al., 2011; Meng et al,. 2007; Nie et al., 2010; Tsang et al., 2008; Vaz et al., 2010). For example, the hypermethylation of *Let-7*a-3 in epithelial ovarian cancers is associated with the low expression of insulin-like growth factor-II and favorable prognosis (Lu et al., 2007).

Producing mature miRNAs goes through different steps which are controlled by different regulatory and transporting machineries for the processing of primary, precursor and mature miRNAs. Finally, the miRNA-associated protein-RNA-induced silencing complex (miRISC) interacts with miRNAs and performs endogenous functions (Meister et al., 2004; Siomi and Siomi, 2010). Recent findings demonstrated that epigenetic aberrations (Heller et al., 2010; Lujambio et al., 2008), miRNA processing and degradation defects also affected the expression of miRNA (Michlewski et al., 2010; Michlewski et al., 2008; Siomi and Siomi, 2010). Therefore, genetic and epigenetic defects likely affect both coding and noncoding RNA transcription, contributing independently or in concert to the altered expression of miRNAs. In ES cells, the core transcription factors, OCT4/SOX2/NANOG/TCF3, promote the transcription of both primary pri-*Let-7*g and LIN28. Expression of pri-*Let-7*g can be detected, but, mature form of *Let-7*g is missing after the blockade of LIN28 (Marson et al., 2008). In this post-transcriptional regulation, two major factors, Drosha and Dicer, are shown to be mediated by LIN28 resulting in the inhibition of *Let-7* maturation (Heo et al., 2008; Rybak et al., 2009; Viswanathan et al., 2008; Zhong et al., 2010). It is suggested that the inhibition of maturation process of pri-*Let-7* transcripts plays an important role in the maintenance of pluripotent state. It is also shown that a feedback loop between LIN28 and *Let-7* regulate pre-*Let-7* maturation during neural stem-cell commitment (Rybak et al., 2009). Thus, the negative LIN28/*Let-7* circuit loop has a major influence over ES cell fate. In addition, LIN28 has been shown as a key player in the effective production of OCT4 and ES cell proliferation (Peng et al., 2010; Qiu et al., 2010). Although, many anti-tumor effects of *Let-7* has been observed but further studies are required to understand the complex and unclear regulatory network of *Let-7* and its target in context of stem/progenitor cells for the self-renewal or differentiation.

2) Current Strategy and Delivery System for the *Let-7*-Based Gene Therapy and Their Limitation

Most recently, many researchers used numbers of strategies to deliver *Let-7* into different cancer cells in experimental models for the purpose of gene therapy. However, clinically applicable tools for the gene delivery are limited and focused on retroviral-based or adenoviral-based vectors. New delivery methods are required to develop for improving the safety and efficacy of *Let-7* gene-based therapy. The first case of *Let-7* functioning as a tumor suppressor *in vivo* has been shown to suppress non-small cell lung tumorigenesis (Kumar et al., 2008). Although the sustained expression of *Let-7* is sufficient to inhibit tumorigenesis, authors infer that *Let-7*g could be present and active in escaping tumors (Kumar et al., 2008). It is suggested that *Let-7* resistant tumors might eventually relapse. It is important to understand the effect of long-term expressed *Let-7* on tumors or other normal cells or the use of *Let-7* miRNAs as a therapeutic agent. A general drawback for current gene therapy strategies is lack of specific marker to target tumor cells while systematically administrating the viral vector carrying the gene of interest. It is critical in order to minimize unexpected side effects. The alternative way for most clinical trials has relied on gene delivery directly into accessible tumors (Roth and Grammer, 2003).

In addition to viral vector-based gene therapy, synthetic miRNA has also been used to in the gain-of-function assays. These miRNA mimics are small, chemically modified RNA molecules that mimic endogenous mature miRNA molecules, and are commercially available (Akinc et al., 2008). By use of the systemic and lipid-based delivery of synthetic miRNA *in vivo*, Wiggins et al. (2010) has successfully shown the therapeutic effect of tumor suppressor miRNA-34 in the lung cancer xenograft model. Since miRNA mimics have no vector-based toxicity, if their delivery agents do not cause side effects over long-term use, it can be a promising therapeutic approach for treatment of cancer.

5. Alternative Ways to Bypass Disadvantages of *Let-7*-Based Therapy

Although it is understandable that *Let-7* plays a critical role in regulating the self-renewal and pluripotency of ES cells, much less is known about the molecular mechanisms for regulating *Let-7* expression to keep the pluripotency in stem/progenitor cells. It is necessary to perform the systematic examination and study of both functional trans-molecules, and cis-regulatory elements involved in the regulation of *Let-7* expression between the transition of stem/progenitor and differentiated cells. A recent study involving large scale identification of *Let-7* promoters in both human and mouse cells is an excellent starting point for characterizing individual miRNA promoters (Marson et al., 2008). Studies in cluster promoters of miRNAs have been demonstrated the binding of SOX2, OCT4, and NANOG in the *Let-7* promoter region. It suggests that the different hierarchical regulation of *Let-7* via biogenesis, integrated circuits with *Let-7* and other factors forms a regulatory feedback loop. As noted above, c-MYC has been shown to be transcriptionally repressing the expression of *Let-7* in lymphoma cells (Chang et al., 2008). Unexpectedly, although mature *Let-7* family members are depleted in undifferentiated cells, the primary *Let-7* transcripts and the hairpin precursors can be detected in these cells *in vitro*, as well as during early developmental stages of mice *in vivo* (Heo et al., 2009; Viswanathan et al., 2010; Yang et al., 2010). The accumulation of mature *Let-7* miRNA thus appears to be regulated, at least in part, post-transcriptionally. Recently, LIN28 and LIN28B, two homologs of the *C. elegans* heterochronic gene LIN28, were shown to inhibit the processing of *Let-7* family members (Newman et al., 2008; Viswanathan et al., 2010; Viswanathan et al., 2008). Additionally, mouse LIN41 has been shown to suppress *Let-7* activity, at least in part, by antagonizing Argonaute 2 (Rybak et al., 2009). Taken together, studies on the expression of *Let-7* have shown a role in inhibiting the induction of pluripotency in progenitor cells. It will be important to study whether depletion of MYC or LIN28 can substitute for the antisense knockdown of *Let-7*. Forced expression of *Let-7* can prevent reprogramming of fibroblasts into iPS cells (Newman et al., 2008; Piskounova et al., 2008). Overexpression of *Let-7* by cotransfection of mature *Let-7* or use of transgenic *Let-7* precursors which are not subjected to LIN28 regulation would be useful in developing *Let-7* based therapies.

6. LIN28 with Multi-Functions, Maybe a Potent Molecule to Block the *Let-7* Associated Tumorigenesis

Inflammation is linked clinically and epidemiologically to cancer, and inappropriate resolution of inflammatory responses often lead to various chronic ailments including cancer (Allavena et al., 2008). In the tumor transformation of breast cells, the transient activation of SRC triggers an inflammatory response mediated by NF-κB that directly activates LIN28 transcription and rapidly reduces *Let-7* levels (Iliopoulos et al., 2010; Ivey et al., 2008). NF-κB is important in regulating normal innate and adaptive immune responses seen in states of inflammation. However, evidence revealing particular mechanisms by which NF-κB influences cancer initiation, promotion, and progression is fascinating and still vague (Naugler and Karin, 2008). It suggests that the activity of *Let-7* is blocked by activation of LIN28 via the epigenetic transforming signaling, NF-κB. Although *Let-7* targets LIN28 expression to block the oncogenetic activity, LIN28 is also involved in a feedback loop to downregulate *Let-7* expression at pri-*Let-7* and pre-*Let-7* processing. Directly elevating *Let-7* in the treatment of cancer therapy is probably inefficient because LIN28 is continuously expressed under the inflammatory microenvironment of tumor site. Because the expression of LIN28 is reciprocal to the maturation of mature *Let-7*, LIN28 is abundant in early developmental stages and declines upon differentiation

Upregulation of *Let-7* expression triggers the differentiation of adult stem/progenitor cell. While using of *Let-7* as a target for the cancer therapy, it has the possibility to impair the replenishment of normal stem/progenitor cell due to non specific tumor-targeting strategy. Suppression of LIN28 may provide an appropriate strategy for the cancer gene therapy. Although LIN28 is found to be highly expressed for maintaining the pluripotency in ES cells by the blockade of *Let-7* expression, LIN28 is shown to differentially promote and inhibit the specific differentiation during neurogliogenesis (Balzer et al., 2010). It suggests that LIN28 does not function exclusively through blocking *Let-7* family. Moreover, in neural stem/progenitor cells, the mechanism for keeping the stemness is associated with the regulation of Musashi1 (MSI1) in concert with LIN28 (Kawahara et al., 2011). This study reveals that Msi1 acts as a novel factor which influencse stem-cell maintenance by controlling synergistically with LIN28 for the blockade of *Let-7* expression.

With all the efforts and advances proceeded to develop miRNA-mediated therapies, obstacles still remain. MiRNA targeting is known to be sequence-specific rather than gene-specific, and *Let-7* family has been shown its multiple targeted genes. It should be noted that silencing genes by miRNAs require partial complementary binding between miRNAs and protein-coding transcripts. Therefore, it is important to evaluate the effect of specific miRNA-mediated therapy on a proteome-wide scale to prevent unwanted gene alteration in non-tumor cells.

Conclusion

In summary, *Let-7* controls multiple targets to regulate stemness and differentiation as required for proper development and tumor suppression. The identity of these targets, and their physiological relevance, has initially emerged. For the future cancer therapeutic, patents for the *Let-7* based therapy in Australia (2007/333109 A1) and USA (20090163430) were filed.

We are now beginning to learn how to regulate the expression of *Let-7* and to find out the tumor-specific genes targeted by *Let-7* for the future *Let-7* based therapy. Such investigation will be valuable in helping us to understand the regulation of cell fates, in particular stem/progenitor cell fates. Therefore, it will be necessary to define the transcriptional regulatory networks and cell signaling pathways that distinguish normal stem/progenitor cells from transformed cells. Although studies show a promising therapeutic approach for treating various kinds of cancer by *Let-7*, it should be noted that these studies looked at the impairment of tumor initiation, not remission of pre-existing tumors, the latter being much associated with the clinically relevant situation. Moreover, over time, tumors developed resistance to sustained expression of *Let-7*g and relapsed (Kumar et al., 2008). Other targets such as HMGA2 or LIN28 might provide the profound regulation of stemness and differentiation to substitute for *Let-7* based cancer therapy.

At the present time, many questions about *Let-7* function in normal stemness, development and differentiation of stem/progenitor cells are still unanswered. The mechanisms of *Let-7* resistance might be emerged during the extended treatment of exogenous *Let-7*. Mostly, additional pluripotent factors or inhibitors for substituting *Let-7* family are continuously addressed in promoting the generation of inducible-Pluripotent Stem cells or cancer therapies.

Acknowledgments

This work was supported by grants from National Science Council (98-2314-B-039-008-MY2 and 100-2314-B-039-005), Taiwan, Republic of China. This study was also supported in part by grants from China Medical University and Hospital (DMR-99-169) and Taiwan Department of Health Clinical Trial and Research Center of Excellence on Stroke and Neurological Diseases (DOH101-TD-B-111-004).

References

Ahn, H. W., Morin, R. D., Zhao, H., Harris, R. A., Coarfa, C., Chen, Z. J., Milosavljevic, A., Marra, M. A., Rajkovic, A. MicroRNA transcriptome in the newborn mouse ovaries determined by massive parallel sequencing. *Mol. Hum. Reprod.* 2010;16:463-471.

Akinc, A., Zumbuehl, A., Goldberg, M., Leshchiner, E. S., Busini, V., Hossain, N., Bacallado, S. A., Nguyen, D. N., Fuller, J., Alvarez, R. et al. A combinatorial library of lipid-like materials for delivery of RNAi therapeutics. *Nat. Biotechnol.* 2008;26:561-569.

Allavena, P., Garlanda, C., Borrello, M. G., Sica, A., Mantovani, A. Pathways connecting inflammation and cancer. *Curr. Opin. Genet. Dev.* 2008;18:3-10.

Almanza, G., Fernandez, A., Volinia, S., Cortez-Gonzalez, X., Croce, C. M., Zanetti, M. Selected microRNAs define cell fate determination of murine central memory CD8 T cells. *PLoS One*. 2010;5:e11243.

Balzer, E., Heine, C., Jiang, Q., Lee, V. M., Moss, E. G. LIN28 alters cell fate succession and acts independently of the *Let-7* microRNA during neurogliogenesis *in vitro*. *Development*. 2010;137:891-900.

Boyerinas, B., Park, S. M., Hau, A., Murmann, A. E., Peter, M. E. The role of *Let-7* in cell differentiation and cancer. *Endocr. Relat. Cancer*. 2010;17:F19-36.

Boyerinas, B., Park, S. M., Shomron, N., Hedegaard, M. M., Vinther, J., Andersen, J. S., Feig, CXuJ, Burge, C. B., Peter, M. E. Identification of *Let-7*-regulated oncofetal genes. *Cancer Res.* 2008;68:2587-2591.

Bussing, I., Slack, F. J., Grosshans, H. *Let-7* microRNAs in development stem cells and cancer. *Trends Mol. Med.* 2008;14:400-409.

Cavazzana-Calvo, M., Payen, E., Negre, O., Wang, G., Hehir, K., Fusil, F., Down, J., Denaro, M., Brady, T., Westerman, K. et al. Transfusion independence and HMGA2 activation after gene therapy of human beta-thalassaemia. *Nature*. 2010;467:318-322.

Caygill, E. E., Johnston, L. A. Temporal regulation of metamorphic processes in Drosophila by the *Let-7* and miR-125 heterochronic microRNAs. *Curr. Biol.* 2008;18:943-950.

Chang, T. C., Yu, D., Lee, Y. S., Wentzel, E. A., Arking, D. E., West, K. M., Dang, C. V., Thomas-Tikhonenko, A., Mendell, J. T. Widespread microRNA repression by Myc contributes to tumorigenesis. *Nat. Genet.* 2008;40:43-50.

Dahiya, N., Morin, P. J. MicroRNAs in ovarian carcinomas. *Endocrine Related Cancer*.2009;17:F77-F89.

Drummond, M. J., McCarthy, J. J., Sinha, M., Spratt, H. M., Volpi, E., Esser, K. A., Rasmussen, B. B. Aging and MicroRNA Expression in Human Skeletal Muscle: A Microarray and Bioinformatics Analysis. *Physiol. Genomics*. 2011;43:595-603.

Gehrke, S., Imai, Y., Sokol, N., Lu, B. Pathogenic LRRK2 negatively regulates microRNA-mediated translational repression. *Nature*. 2010;466:637-641.

Heller, G., Zielinski, C. C., Zochbauer-Muller, S. Lung cancer: from single-gene methylation to methylome profiling. *Cancer Metastasis Rev*. 2010;29:95-107.

Heo, I., Joo, C., Cho, J., Ha, M., Han, J., Kim, V. N. LIN28 mediates the terminal uridylation of *Let-7* precursor. *MicroRNA Mol. Cell*. 2008;32:276-284.

Heo, I., Joo, C., Kim, Y. K., Ha, M., Yoon, M. J., Cho, J., Yeom, K. H., Han, J., Kim, V. N. TUT4 in concert with LIN28 suppresses microRNA biogenesis through pre-microRNA uridylation. *Cell*. 2009;138:696-708.

Hermann PC, Bhaskar S. Cioffi M Heeschen C. Cancer stem cells in solid tumors. *Semin. Cancer Biol.* 2010;20:77-84.

Ibarra, I., Erlich, Y., Muthuswamy, S. K., Sachidanandam, R., Hannon, G. J. A role for microRNAs in maintenance of mouse mammary epithelial progenitor cells. *Genes Dev.* 2007;21:3238-3243.

Iliopoulos, D., Hirsch, H. A., Struhl, K. An epigenetic switch involving NF-kappaB LIN28 *Let-7* MicroRNA and IL6 links inflammation to cell transformation. *Cell.* 2009;139:693-706.

Iliopoulos, D., Jaeger, S. A., Hirsch, H. A., Bulyk, M. L., Struhl, K. STAT3 activation of miR-21 and miR-181b-1 via PTEN and CYLD are part of the epigenetic switch linking inflammation to cancer. *Mol. Cell.*.2010;39:493-506.

Ivey, K. N., Muth, A., Arnold, J., King, F. W., Yeh, R. F., Fish, J. E., Hsiao, E. C., Schwartz, R. J., Conklin, B. R., Bernstein, H. S. et al. MicroRNA regulation of cell lineages in mouse and human embryonic stem cells. *Cell Stem. Cell.* 2008;2:219-229.

Johnson, C. D., Esquela-Kerscher, A., Stefani, G., Byrom, M., Kelnar, K., Ovcharenko, D., Wilson, M., Wang, X., Shelton, J., Shingara, J. et al. The *Let-7* microRNA represses cell proliferation pathways in human cells. *Cancer Res.* 2007;67:7713-7722.

Jung, Y. H., Gupta, M. K., Shin, J. Y., Uhm, S. J., Lee, H. T. MicroRNA signature in testes-derived male germ-line stem cells. *Mol. Hum. Reprod.* 2010;16:804-810.

Kawahara, H., Okada, Y., Imai, T., Iwanami, A., Mischel, P. S., Okano, H. Musashi 1 cooperates in abnormal cell LINeage protein 28 (LIN28)-mediated *Let-7* family microRNA biogenesis in early neural differentiation. *J. Biol. Chem.* 2011;286:16121-16130.

Koh, W., Sheng, C. T., Tan, B., Lee, Q. Y., Kuznetsov, V., Kiang, L. S., Tanavde, V. Analysis of deep sequencing microRNA expression profile from human embryonic stem cells derived mesenchymal stem cells reveals possible role of *Let-7* microRNA family in downstream targeting of hepatic nuclear factor 4 alpha. *BMC Genomics*. 2010;11:S1:S6.

Kuehbacher, A., Urbich, C., Zeiher, A. M., Dimmeler, S. Role of Dicer and Drosha for endothelial microRNA expression and angiogenesis. *Circ. Res.* 2007;101:59-68.

Kumar, M. S., Erkeland, S. J., Pester, R. E., Chen, C. Y., Ebert, M. S., Sharp, P. A., Jacks, T. Suppression of non-small cell lung tumor development by the *Let-7* microRNA family *Proc. Natl. Acad. Sci.* U S A. 2008;105:3903-3908.

Lee, Y. S., Dutta, A. The tumor suppressor microRNA *Let-7* represses the HMGA2 oncogene. *Genes Dev.* 2007;21:1025-1030.

Li, G., Li, Y., Li, X., Ning, X., Li, M., Yang, G. MicroRNA identity and abundance in developing swine adipose tissue as determined by Solexa sequencing. *J. Cell Biochem.* 2011.

Lu, L., Katsaros, D., De la Longrais, I. A., Sochirca, O., Yu, H. Hypermethylation of *Let-7*a-3 in epithelial ovarian cancer is associated with low insulin-like growth factor-II expression and favorable prognosis. *Cancer Res.* 2007;67:10117-10122.

Lujambio, A., Calin, G. A., Villanueva, A., Ropero, S., Sanchez-Cespedes, M., Blanco, D., Montuenga, L. M., Rossi, S., Nicoloso, M. S., Faller, W. J. et al. A microRNA DNA methylation signature for human cancer metastasis. *Proc. Natl. Acad. Sci.* U S A. 2008;105:13556-13561.

Marson, A., Levine, S. S., Cole, M. F., Frampton, G. M., Brambrink, T. Johnstone, S. Guenther, M. G., Johnston, W. K., Wernig, M., Newman, J., and others Connecting microRNA genes to the core transcriptional regulatory circuitry of embryonic stem cells. *Cell.* 2008;134:521-533.

Meister, G., Tuschl, T. Mechanisms of gene silencing by double-stranded RNA. *Nature.* 2004;431:343-349.

Melton, C., Judson, R. L., Blelloch, R. Opposing microRNA families regulate self-renewal in mouse embryonic stem cells. *Nature.* 2010:463:621-626.

Meng, F., Glaser, S. S., Francis, H., Demorrow, S., Han, Y., Passarini, J. D., Stokes, A., Cleary, J. P., Liu, X., Venter, J. et al. Functional Analysis of microRNAs in Human Hepatocellular Cancer Stem Cells. *J. Cell Mol. Med.* 2011.

Meng, F., Henson, R., Wehbe-Janek, H., Smith, H., Ueno, Y., Patel, T. The MicroRNA *Let-7*a modulates interleukin-6-dependent STAT-3 survival signaling in malignant human cholangiocytes. *J. Biol. Chem.* 2007;282:8256-8264.

Michlewski, G., Caceres, J. F. Antagonistic role of hnRNP A1 and KSRP in the regulation of *Let-7*a biogenesis. *Nat. Struct. Mol. Biol.* 2010;17:1011-1018.

Michlewski, G., Guil, S., Semple, C. A., Caceres, J. F. Posttranscriptional regulation of miRNAs harboring conserved terminal loops. *Mol. Cell.* 2008;32:383-393.

Mongroo, P. S., Noubissi, F. K., Cuatrecasas, M., Kalabis, J., King, C. E., Johnstone, C. N., Bowser, M. J., Castells, A., Spiegelman, V. S., Rustgi, A.K. IMP-1 displays crosstalk with K-Ras and modulates colon cancer cell survival through the novel pro-apoptotic protein CYFIP2. *Cancer Res.* 2011.

Nakamura, K., Maki, N., Trinh, A., Trask, H. W., Gui, J., Tomlinson, C. R., Tsonis, P. A. miRNAs in newt lens regeneration: specific control of proliferation and evidence for miRNA networking. *PLoS One.* 2010;5:e12058.

Naugler, W. E., Karin, M. NF-kappaB and cancer-identifying targets and mechanisms. *Curr. Opin. Genet. Dev.* 2008;18:19-26.

Newman, M. A., Thomson, J. M., Hammond, S. M. LIN28 interaction with the *Let-7* precursor loop mediates regulated microRNA processing. *RNA.* 2008;14:1539-1549.

Nie, K., Zhang, T., Allawi, H., Gomez, M., Liu, Y., Chadburn, A., Wang, Y. L., Knowles, D. M., Tam, W. Epigenetic down-regulation of the tumor suppressor gene PRDM1/Blimp-1 in diffuse large B cell lymphomas: a potential role of the microRNA *Let-7. Am. J. Pathol.* 2010;177:1470-1479.

Nilsen, T. W. Mechanisms of microRNA-mediated gene regulation in animal cells. *Trends Genet.* 2007;23:243-249.

Nishino, J., Kim, I., Chada, K., Morrison, S. J. Hmga2 promotes neural stem cell self-renewal in young but not old mice by reducing p16Ink4a and p19Arf Expression. *Cell.* 2008;135:227-239.

Noh, S. J., Miller, S. H., Lee, Y. T., Goh, S. H., Marincola, F. M., Stroncek, D. F., Reed, C., Wang, E., Miller, J. L. *Let-7* microRNAs are developmentally regulated in circulating human erythroid cells. *J. Transl. Med.* 2009;7:98.

O'Day, E., Lal, A. MicroRNAs and their target gene networks in breast cancer *Breast Cancer Res.* 2010;12:201.

Ohshima, K., Inoue, K., Fujiwara, A., Hatakeyama, K., Kanto, K., Watanabe, Y., Muramatsu, K., Fukuda, Y., Ogura, S., Yamaguchi, K. et al. *Let-7* microRNA family is selectively secreted into the extracellular environment via exosomes in a metastatic gastric cancer cell line. *PLoS One.* 2010;5:e13247.

Ortholan, C., Puissegur, M. P., Ilie, M., Barbry, P., Mari, B., Hofman, P. MicroRNAs and lung cancer: new oncogenes and tumor suppressors new prognostic factors and potential therapeutic targets. *Curr. Med. Chem.* 2009;16:1047-1061.

Pasquinelli, A. E., Reinhart, B. J., Slack, F., Martindale, M. Q., Kuroda, M. I., Maller, B., Hayward, D. C., Ball, E. E., Degnan, B., Muller, P. et al. Conservation of the sequence and temporal expression of *Let-7* heterochronic regulatory RNA. *Nature.* 2000;408:86-89.

Peng, S., Maihle, N. J., Huang, Y. Pluripotency factors LIN28 and Oct.4 identify a sub-population of stem cell-like cells in ovarian cancer. *Oncogene.* 2010;29:2153-2159.

Piskounova, E., Viswanathan, S. R., Janas, M., LaPierre, R. J., Daley, G. Q., Sliz, P., Gregory, R. I. Determinants of microRNA processing inhibition by the developmentally regulated RNA-binding protein LIN28. *J. Biol. Chem.* 2008;283:21310-21314.

Polikepahad, S., Knight, J. M., Naghavi, A. O., Oplt, T., Creighton, C. J., Shaw, C., Benham, A. L., Kim, J., Soibam, B., Harris, R. A. et al. Proinflammatory role for *Let-7* microRNAs in experimental asthma. *J. Biol. Chem.* 2010;285:30139-30149.

Qiu, C., Ma, Y., Wang, J., Peng, S., Huang, Y. LIN28-mediated post-transcriptional regulation of Oct4 expression in human embryonic stem cells. *Nucleic Acids Res.* 2010;38:1240-1248.

Ramachandran, R., Fausett, B. V., Goldman, D. Ascl1a regulates Muller glia dedifferentiation and retinal regeneration through a LIN28-dependent *Let-7* microRNA signalling pathway. *Nat. Cell Biol.* 2010;12:1101-1107.

Reinhart, B. J., Slack, F. J., Basson, M., Pasquinelli, A. E., Bettinger, J. C., Rougvie, A. E., Horvitz, H. R., Ruvkun, G. The 21-nucleotide *Let-7* RNA regulates developmental timing in Caenorhabditis elegans. *Nature.* 2000;403:901-906.

Roth, J. A., Grammer, S. F. Tumor suppressor gene therapy. *Methods Mol. Biol.* 2003;223:577-598.

Roush, S., Slack, F. J. The *Let-7* family of microRNAs. *Trends Cell Biol.* 2008;18:505-516.

Rybak, A., Fuchs, H., Hadian, K., Smirnova, L., Wulczyn, E. A., Michel, G., Nitsch, R., Krappmann, D., Wulczyn, F. G. The *Let-7* target gene mouse lin-41 is a stem cell specific E3 ubiquitin ligase for the miRNA pathway protein Ago2. *Nat. Cell Biol.* 2009;11:1411-1420.

Rybak, A., Fuchs, H., Smirnova, L., Brandt, C., Pohl, E. E., Nitsch, R., Wulczyn, F. G. A feedback loop comprising LIN28 and *Let-7* controls pre-*Let-7* maturation during neural stem-cell commitment. *Nat. Cell Biol.* 2008;10:987-993.

Schultz, J., Lorenz, P., Gross, G., Ibrahim, S., Kunz, M. MicroRNA *Let-7*b targets important cell cycle molecules in malignant melanoma cells and interferes with anchorage-independent growth. *Cell Res.* 2008;18:549-557.

Schwamborn, J. C., Berezikov, E., Knoblich, J. A. The TRIM-NHL protein TRIM32 activates microRNAs and prevents self-renewal in mouse neural progenitors. *Cell.* 2009;136:913-925.

Sempere, L. F., Freemantle, S., Pitha-Rowe, I., Moss, E., Dmitrovsky, E., Ambros, V. Expression profiling of mammalian microRNAs uncovers a subset of brain-expressed microRNAs with possible roles in murine and human neuronal differentiation. *Genome Biol.* 2004;5:R13.

Siomi, H., Siomi, M. C. Posttranscriptional regulation of microRNA biogenesis in animals. *Mol. Cell.* 2010;38:323-332.

Sokol, N. S., Xu, P., Jan, Y. N., Ambros, V. Drosophila *Let-7* microRNA is required for remodeling of the neuromusculature during metamorphosis. *Genes Dev.* 2008;22:1591-1596.

Stefani, G., Slack, F. J. Small non-coding RNAs in animal development. *Nat. Rev. Mol. Cell Biol.* 2008;9:219-230.

Sun, T., Fu, M., Bookout, A. L., Kliewer, S. A., Mangelsdorf, D. J. MicroRNA *Let-7* regulates 3T3-L1 adipogenesis. *Mol. Endocrinol.* 2009;23:925-931.

Tsang, W. P., Kwok, T. T. *Let-7*a microRNA suppresses therapeutics-induced cancer cell death by targeting caspase-3. *Apoptosis.* 2008;13:1215-1222.

Tsonis, P. A., Call, M. K., Grogg, M. W., Sartor, M. A., Taylor, R. R., Forge, A., Fyffe, R., Goldenberg, R., Cowper-Sal-lari, R., Tomlinson, C. R. MicroRNAs and regeneration: *Let-7* members as potential regulators of dedifferentiation in lens and inner ear hair cell regeneration of the adult newt. *Biochem. Biophys. Res. Commun.* 2007;362:940-945.

Tzur, G., Israel, A., Levy, A., Benjamin, H., Meiri, E., Shufaro, Y., Meir, K., Khvalevsky, E., Spector, Y., Rojansky, N. et al. Comprehensive gene and microRNA expression profiling reveals a role for microRNAs in human liver development. *PLoS One.* 2009;4:e7511.

Van Wynsberghe, P. M., Kai, Z. S., Massirer, K. B., Burton, V. H., Yeo, G. W., Pasquinelli, A. E. LIN28 co-transcriptionally binds primary *Let-7* to regulate miRNA maturation in Caenorhabditis elegans. *Nat. Struct. Mol. Biol.* 2011;18:302-308.

Vaz, C., Ahmad, H. M., Sharma, P., Gupta, R., Kumar, L., Kulshreshtha, R., Bhattacharya, A. Analysis of microRNA transcriptome by deep sequencing of small RNA libraries of peripheral blood. *BMC Genomics.* 2010;11:288.

Viswanathan, S. R., Daley, G. Q. LIN28: A microRNA regulator with a macro role. *Cell.* 2010;140:445-449.

Viswanathan, S. R., Daley, G. Q., Gregory, R. I. Selective blockade of microRNA processing by LIN28. *Science.* 2008;320:97-100.

Wang, S., Tang, Y., Cui, H., Zhao, X., Luo, X., Pan, W., Huang, X., Shen, N. *Let-7*/miR-98 regulate Fas and Fas-mediated apoptosis. *Genes Immun.* 2011;12:149-154.

West, J. A., Viswanathan, S. R., Yabuuchi, A., Cunniff, K., Takeuchi, A., Park, I. H., Sero, J. E., Zhu, H., Perez-Atayde, A., Frazier, A. L. et al. A role for LIN28 in primordial germ-cell development and germ-cell malignancy. *Nature.* 2009;460:909-913.

Wiggins, J. F., Ruffino, L., Kelnar, K., Omotola, M., Patrawala, L., Brown, D Bader AG. Development of a lung cancer therapeutic based on the tumor suppressor microRNA-34. *Cancer Res.* 2010;70:5923-5930.

Wulczyn, F. G., Smirnova, L., Rybak, A., Brandt, C., Kwidzinski, E., Ninnemann, O., Strehle, M., Seiler, A., Schumacher, S., Nitsch, R. Post-transcriptional regulation of the *Let-7* microRNA during neural cell specification. *FASEB J.* 2007;21:415-426.

Yang, X., Lin, X., Zhong, X., Kaur, S., Li, N., Liang, S., Lassus, H., Wang, L., Katsaros, D., Montone, K. et al. Double-negative feedback loop between reprogramming factor LIN28 and microRNA *Let-7* regulates aldehyde dehydrogenase 1-positive cancer stem cells. *Cancer Res.* 2010;70:9463-9472.

Yang, Y., Ago, T., Zhai, P., Abdellatif, M., Sadoshima, J. Thioredoxin 1 negatively regulates angiotensin II-induced cardiac hypertrophy through upregulation of miR-98/*Let-7*. *Circ. Res.* 2011;108:305-313.

Zhao, C., Sun, G., Li, S., Lang, M. F., Yang, S., Li, W., Shi, Y., MicroRNA *Let-7*b regulates neural stem cell proliferation and differentiation by targeting nuclear receptor TLX signaling. *Proc. Natl. Acad. Sci.* U S A. 2010;107:1876-1881.

Zhong, X., Li, N., Liang, S., Huang, Q., Coukos, G., Zhang, L. Identification of microRNAs regulating reprogramming factor LIN28 in embryonic stem cells and cancer cells. *J. Biol Chem.* 2010;285:41961-41971.

Zhu, H., Shah, S., Shyh-Chang, N., Shinoda, G., Einhorn, W. S., Viswanathan, S. R., Takeuchi, A., Grasemann, C., Rinn, J. L., Lopez, M. F. et al. LIN28a transgenic mice manifest size and puberty phenotypes identified in human genetic association studies. *Nat. Genet.* 2010;42:626-630.

In: MicroRNA *Let-7*
Editor: Neetu Dahiya

ISBN: 978-1-62081-152-8

Chapter 6

THE IMPACT OF THE *LET-7* FAMILY ON RESISTANCE TO ANTICANCER TREATMENT

M. Fritz,[1] D. J. Hussey,[3] J. Haier[2] and R. Hummel[1,*]

[1]Department of General and Visceral Surgery,
University of Muenster, Germany
[2]Comprehensive Cancer Center, University of Muenster, Germany
[3]Department of Surgery, Flinders University Adelaide, Australia

1. ABSTRACT

MiRNAs (miRNAs) represent a class of naturally occurring, small (19 to 25-nucleotides), non-coding RNA molecules. Among 16000 miRNA genes which have been identified in various species, more than 1,000 miRNAs have been reported in the humans. The lethal-7 (*Let-7*) gene family, which was initially found as an essential developmental gene family in *Caenorhabditis elegans,* is one of the most extensively studied miRNA families, and there is high evidence that these miRNAs play key roles in the initiation, growth and progression of a variety of (malignant) tumors. Based on these data, there is now a growing interest in using these molecules for clinical purposes. The present chapter therefore discusses and highlights potential diagnostic applications of members of the *Let-7* family as predictors of clinical outcome and as markers for response monitoring during chemo- and/ or radiotherapy. In addition, we present first promising data on the *Let-7* family concerning their use as primary anticancer agents and, perhaps more importantly, as modifiers and enhancers of well-established anticancer therapy strategies. In this context, modulation of *Let-7* expression seems to provide a new and very promising first-line treatment option in cancer, as well as an additive therapeutic modality to chemo- and radiotherapy, augmenting their effects and possibly reversing drug resistance – a major obstacle in modern anticancer treatment.

* Corresponding author: R. Hummel. Department of General and Visceral Surgery, University Hospital of Muenster Waldeyerstr 1, 48149 Muenster, Germany. E-Mail: richard.hummel@ukmuenster.de.

2. Introduction

MiRNAs (miRNAs) are components of the cellular 'epigenetic machinery', which has a profound effect on the modulation of gene expression (Mulero-Navarro and Esteller, 2008). Today more than 16000 miRNA genes have been identified in various species, with over 1000 of these miRNAs in humans (Griffiths-Jones et al., 2008). MiRNAs biogenesis has been examined by many authors (Garzon et al., 2006; Zhang et al., 2007, Bartel, 2004). Several biological functions such as organ morphology, stress response, metabolism, cell proliferation and apoptosis are regulated by these small RNA molecules (Ambros, 2003). In cancer, they play a crucial role as onco-miRs and tumor suppressor-miRs (Pramanik et al., 2011; Hummel et al., 2010).

The *Let-7* gene was identified as a developmental gene in *Caenorhabditis elegans*. It was discovered as the second miRNA after lin-4 in this species (Reinhart et al., 2000), and is now one of the most extensively studied miRNAs. The human *Let-7* family contains 13 members located on nine different chromosomes. The nomenclature differentiates between the multiple isoforms by placing a letter after "*Let-7*" to indicate an isoform with a slightly different sequence. In addition, a number at the end denotes that the same sequence is present in multiple genomic locations. However, the number of isoforms of this miRNA family differs between species. The genome of *Drosophila* (fly) contains only one *Let-7*, whereas the zebrafish *Danio rerio* has 11 mature sequences in 19 loci throughout its genome (Roush and Slack, 2008).

Biological functions of *Let-7* include the regulation of stem-cell differentiation, neuromusculature and limb development, cell proliferation and differentiation. Moreover, altered expression levels of various members of the *Let-7*-family were found in a variety of malignancies such as in lung, stomach, colon and esophageal cancer (Qing et al., 2011; Guan, et al. 2011; Tsujiura et al., 2010; Akao et al., 2006), and there is now increasing evidence that this miRNA seems to have a profound effect on tumor progression or disease outcome by acting mainly as a tumor suppressor in cancer. Low *Let-7* expression for example has been found to be associated with poor survival of cancer patients (Takamizawa et al., 2004; Yang et al., 2008), or with low tumor differentiation (Shell et al., 2007).

With this background, this book chapter is focussed on the clinical importance of the *Let-7* family and on the possible use of this miRNA in clinical settings. In this context, first reports highlighted that members of the *Let-7* family might impact the response of various tumors to anticancer treatment. As this aspect of *Let-7* (in combination with the ability to predict clinical outcome) has imminent potential as a very promising new approach for the clinical use of these little molecules, our current book chapter aims to provide an overview about these specific characteristics of *Let-7*. In the first part of the manuscript, we focus on a potential diagnostic application of *Let-7* as predictor of clinical outcome and response to anticancer treatment, and on the potential of *Let-7* as marker to monitor success of chemotherapy or irradiation. In the second part of the article, we then present the most recent data about *Let-7* as a primary anticancer treatment option or, even more important, as a modifier of well-established anticancer therapy strategies.

3. Methods

We performed a PubMed search with various combinations of the following keywords: chemotherapy, radiotherapy, irradiation, drug, treatment, resistance, response and *Let-7*. The following numbers of articles were identified to meet the criteria: *Let-7* and chemotherapy (Lee et al., 2009), *Let-7* and radiotherapy (Zhang et al., 2007), *Let-7* and irradiation (Zhang et al., 2007), *Let-7* and drug (Childs et al., 2009), *Let-7* and treatment (Zhang et al., 2009), *Let-7* and resistance (Lu et al., 2007), *Let-7* and response (Wong et al., 2011). From these articles relevant publications dealing with *Let-7* family miRNAs and their impact on chemo- and radiosensitivity were obtained. Further, relevant articles were extracted by screening the references of these papers. In case of non-availability of the whole article the abstract was taken into consideration despite the limited data provided.

4. *Let-7* as a Diagnostic Tool

The use of molecular markers such as *Let-7* as diagnostic tools requires that these markers present a different expression pattern in patients who present either good or poor clinical outcome with respect to disease risk, metastasis or survival. More importantly, a potential use as monitoring tool or predictor of anticancer therapies mandates an altered expression of this marker for example after exposure to various treatments, or an altered expression between resistant and sensitive tumors. And indeed, there is a growing body of evidence that *Let-7* fulfils these criteria and could therefore be used as a diagnostic tool.

1) *Let-7*: Predictor of Clinical Outcome in Cancer

There is increasing evidence that expression profiles of members of the *Let-7*-family correlate with the clinical outcome in a variety of cancers. In colon cancer for example, several studies demonstrated that *Let-7* has an effect as tumor suppressor by targeting the KRAS and binding to the 3′-UTR of the respective mRNA. (Johnson et al., 2005) Zhang and colleagues could show that a polymorphism in a *Let-7* miRNA complementary site (LCS6) in this region was associated with an increased cancer risk, but they found no associations between KRAS *Let-7* LCS6 polymorphism and overall survival or progression-free survival either in the wild-type or in mutant KRAS patients (Zhang et al., 2011). Graziano et al. (2010) on the other hand reported a lower rate of overall survival and progression-free survival when comparing carriers of the LCS6 G-allele genotype and carriers of the wild-type T/T genotype. In addition, several studies have shown an association between LCS6 variant allele and an increased cancer risk in non-small-cell lung cancer among moderate smokers, or a reduced overall survival in oral cavity carcinoma. (Chin et al., 2008; Christensen et al., 2009; Takamizawa et al., 2004) This data suggests that LCS6 polymorphism may impact the patient outcome and survival rate. In another tumor entity, pancreatic ductal adenocarcinoma, *Let-7* levels showed prognostic significance. Expression levels were strongly reduced in pancreatic ductal adenocarcinoma samples obtained after surgery or from fine needle aspiration. It was confirmed that *Let-7* was present in normal pancreatic cells, but expression was lost in poorly differentiated cancer samples or, most interestingly, in patients who did not qualify for surgery (Torrisani et al., 2009). As described above, *Let-7* expression correlated inversely with expression of RAS in human lung cancer tissue and was also associated with poor prognosis (Takamizawa et al., 2004) In addition low *Let-7* expression was also associated with poor prognosis in adenocarcinoma and squamous cell carcinoma of non-small cell lung cancer (NSCLC) (Yanaihara et al., 2006; Raponi et al., 2009). Another very interesting aspect is the methylation status of *Let-7*a-3. It was shown that the *Let-7*a-3 gene was heavily methylated in normal human tissues but hypomethylated in some lung adenocarcinomas implicating that *Let-7*a-3 might have oncogenic properties in this context, which is in contrast to most reports of the function of *Let-7* family miRNAs (Brueckner et al. 2007). And indeed, methylation of *Let-7*a-3 gene was demonstrated to be inversely correlated with IGF-II, which is an indicator of poor prognosis in ovarian cancer and associated with increased risk of tumor progression (Lu et al., 2006). Patients with methylated *Let-7*a-3 seemed to have reduced disease risk, independent of patient age at surgery, tumor grade and disease stage (Lu et al., 2007). Another study on primary epithelial ovarian cancer demonstrated that *Let-7* can be blocked by LIN28 or LIN28B which are RNA-binding proteins. LIN28B was inversely correlated with levels of mature *Let-7*a, and it was shown that high expression of LIN28B was associated with the risk of disease progression and death (Lu et al., 2009). Another work addressed the importance of *Let-7* in head and neck cancer (HNC). The authors reported, that the expression levels of *Let-7* family members were reduced in HNC tissue compared to normal tissue, and that oral cancer stemness expression markers (NANOG and OCT4) were higher in HNC tissue. In order to elucidate possible correlations between expression of *Let-7*a and HNC progression, the authors analyzed gene expression in metastatic and recurrent HNC patients. They could show that high expression of *Let-7*a correlated with an early T-stage, low lymph node metastasis, and early pathological stage. Moreover, *Let-7*a expression was significantly reduced in metastatic HNC tissues compared to primary HNC, and low expression of *Let-7*a in HNC was inversely correlated with tumor relapse. Again, NANOG and OCT4 expression levels were also elevated in metastatic and recurrent HNC tissues (Yu et al., 2011). This data was supported by another study investigating *Let-7* expression in patients with primary HNSCC tumors out of a high-risk area of New York City. *Let-7*d showed lower expression levels in tumors compared with normal tissues, and there was a significantly shorter time to cancer progression or death in HNSCC patients with low levels of *Let-7*d (Childs et al., 2009). Ali et al. (2010) compared the expression of miRNAs in patients with pancreatic cancer to healthy volunteers. They showed that expression of several members of the *Let-7* family, especially *Let-7*d and *Let-7*b was significantly reduced in patients with pancreatic cancer. Low levels of *Let-7*d expression in patients with pancreatic cancer appeared to be associated with a lower rate of survival, and the authors concluded that *Let-7* could be useful for the screening of high-risk patients. One of the resistance factors that protect ovarian cancer cells from Taxol-based therapy is multi-drug resistance 1 (MDR1). It is known, that *Let-7*g selectively affects the sensitivity to the drug by targeting IMP-1. Upregulation of MDR-1 and IMP-1 result in an adverse prognosis of patients treated with Taxanes, whereas a reduction of MDR1 renders cells more sensitive to the two MDR1 substrates Taxol and Vinblastine but not to Carboplatin. Boyerinas and colleagues investigated if an increase in the expression of both proteins correlates with the number of disease-free days between the beginning of chemotherapy and the time of recurrence, and the authors could demonstrate that patients, in whom both proteins were upregulated, had a poorer prognosis. (Boyerinas et al., 2011) Furthermore, in epithelial ovarian cancer, low *Let-7*i expression was significantly associated with shorter progression-free survival as compared to the high *Let-7*i expression group. The authors concluded, that *Let-7*i is an important predictor for chemotherapy, and can also serve as a prognostic marker (Yang et al., 2008)

2) *Let-7*: Predictor of Response to Anticancer Treatment

An association between gene expression and response to anticancer treatment is even more clinically important than a simple correlation between *Let-7* expression and prognosis. As described above, several studies have highlighted the fact that *Let-7* acts as tumor suppressor by targeting KRAS. As already pointed out, a polymorphism in a *Let-7* miRNA complementary site (LCS6) in this gene was associated with an increased cancer risk. Interestingly, KRAS *Let-7* LCS6 polymorphism was found to be related to the response to cetuximab monotherapy in patients with metastatic colorectal cancer, whose tumors expressed KRAS wild-type (Zhang et al., 2011). In pancreatic cancer, mesenchymal-type cancer cells were demonstrated to be more resistant to chemotherapeutic agents than epithelial-type cancer cells. Cell line experiments comparing gemcitabine-sensitive and gemcitabine-resistant pancreatic cancer cells showed that many members of the *Let-7* family were downregulated in gemcitabine-resistant cells. Moreover, gemcitabine-resistant pancreatic cancer cells presented an expression of mesenchymal markers. Furthermore, nontoxic “natural agents” such as B-DIM and isoflavone were shown to be able to upregulate the expression of various miRNAs, including *Let-7*, which resulted in the reversal of epithelial-to-mesenchymal transition of gemcitabine-resistant cells (Li et al., 2009). Another very interesting study investigated *Let-7* in epithelial ovarian cancer, which is one of the most frequent cancer types in women. The therapeutic options for this tumor include surgery and platinum-based chemotherapy resulting in an improved survival. However, in a high number of patients tumors relapse with a subsequent resistance to platinum-based chemotherapy. Lu et al. (2011) have shown that *Let-7*a expression was not associated with disease stage, tumor grade, histology or debulking results. However, survival in patients with high *Let-7*a was better compared to those with low *Let-7*a when treated with platinum-based chemotherapy without paclitaxel. Interestingly, in the treatment group with platinum-based chemotherapy and paclitaxel, high *Let-7*a levels were associated with worse progression-free and overall survival. The authors concluded from this data that the beneficial impact of the addition of paclitaxel to chemotherapeutic treatment on survival was significantly associated to *Let-7*a levels, and emphasized the importance of miRNAs such as *Let-7*a as useful biomarkers for selection of chemotherapeutic agents (Lu et al., 2011). Accordingly, Yang et al. (2008) found that *Let-7*i expression was significantly reduced in patients with chemotherapy-resistant ovarian cancer by analyzing specimens from 72 late-stage patients. They showed that 34 miRNAs were statistically deregulated between the complete response and non-complete response groups, with 24 miRNAs showing higher expression in the non-complete response group, and 10 miRNAs showing higher expression in the complete response group. *Let-7*i showed the largest difference in expression between the two groups and was expressed at remarkably lower levels in the non-complete response group. The authors concluded that low *Let-7*i expression increases the chemotherapy resistance of epithelial ovarian cancer cells (Yang et al. 2008). All these findings are summarized in figure 1.

3) *Let-7*: Tool to Monitor Treatment Success

Only a few articles have described the effects of different anticancer therapies on *Let-7* expression, possibly opening a new approach to monitor success of these therapies. One very interesting study showed that expression of various members of the *Let-7* family, except *Let-7*g, decreased significantly two hours after irradiation in lung cancer cells and normal lung epithelium.

*Let-7*a and *Let-7*b for example were significantly downregulated eight hours after irradiation compared to the baseline expression, whereas *Let-7*g expression increased significantly twenty-four hours after irradiation in all lung cell lines studied. The authors concluded that the similar responses to irradiation in multiple cancerous and normal lung epithelial cells suggest the existence of a global miRNA response in lung cells after irradiation, and that miRNAs might be components of the cellular response to the cytotoxic agents. The same study further investigated vulval precursor cells in a *C. elegans* based *in vivo* model. These experiments showed specific expression of three *Let-7* paralogues (*Let-7*, mir-48 and mir-84), which repress RAS expression. On irradiation, vulval precursor cells in strains overexpressing either *Let-7* or mir-84 were significantly more radiosensitive, presenting additional evidence that the *Let-7* family might modulate the response to cytotoxic anticancer therapy (Weidhaas et al., 2007). Chaudhry and co-authors performed studies on glioblastoma cell lines (M059K and M059J) and found that exposure of cells to ionizing radiation (3Gy) led to an upregulation in *Let-7*a, *Let-7*b, *Let-7*c, *Let-7*d, *Let-7*e, and *Let-7*f levels at the 12-hour time point in M059K cells. *Let-7*g and *Let-7*i showed two upregulation peaks in M059K cells after radiation treatment, at the 4-hour time point and at the 12-hour time point. Compared to M059K, there was a downregulation in the 24-hour time period for the M059J cells. The highest level of downregulation was observed at the 8-hour time point.

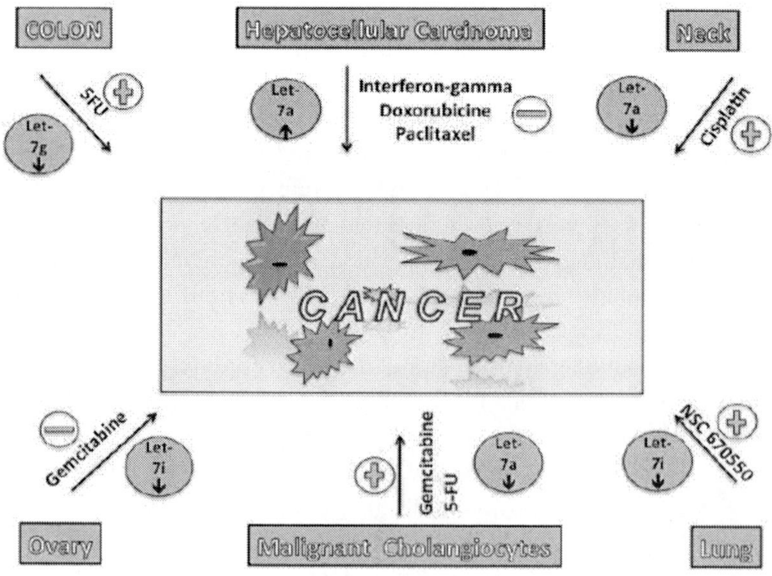

Figure 1. Effects of modulating expression of specific members of the *Let-7* family (arrows under *Let-7*-family member indicate experimental up- or downregulation of the miRNA) on sensitivity to chemotherapy ((+): increased sensitivity; (-): decreased sensitivity).

This study suggests the involvement of members of the *Let-7* family in the response of glioblastoma cells to ionizing radiation, and the authors discussed modulation of RAS oncogenes (which have been shown to increase the intrinsic resistance to ionizing radiation) by *Let-7* as the potentially underlying mechanism. However, these authors point out that the radiation-induced regulation of miRNA expression seems to be far more complex and additional studies are warranted. (Chaudhry, 2008; Chaudhry et al., 2010).

5. *Let-7* as a Therapeutic Tool

Based on the rapidly growing knowledge about miRNAs and their impact on cancer development and progression, it seems only logical to expand their possible clinical applications beyond diagnostic approaches, and to evaluate their potential as new therapeutic options in the fight against cancer. In principal, there are two major ways that miRNAs, and in the context of this chapter: *Let-7*, might be used as therapeutics: on the one hand, anti-proliferative and pro-apoptotic effects of single members of the *Let-7* family could be used, and these molecules could be administered as first line or primary treatment aiming to inhibit tumor growth and induce cell death directly. On the other hand, a far more promising approach might lie in the possibility to support conventional and well-established anticancer therapies such as chemotherapy or irradiation, and to augment the effects of these treatment options and to overcome resistance to various agents. And indeed, not only other miRNAs but also several members of the *Let-7* family seem to impact on tumor cell survival and drug resistance.

1) Impact of *Let-7* on Tumor Cell Survival

Here are some of the evidences showing anti-tumor effect of *Let-7* modulation. In lung cancer for example, recent studies reported that *Let-7* expression levels were reduced in various lung cancer cell lines and pulmonary tumors. In experimental settings with xenograft models, Esquela-Kerscher et al. (2008) demonstrated that *Let-7*b inhibited the tumor growth of the A549 lung cancer cell lines. In pancreatic cancer, *Let-7* transfection caused an inhibition of cell proliferation, K-RAS expression and MAPK activation in capan-1 cells derived from human pancreatic ductal adenocarcinoma (Torrisani et al., 2009). Another important aspect is the fact that *Let-7* targets HMGA2. Studies showed that HMGA2 is more efficiently targeted than RAS by *Let-7*. HMGA2 expression was inversely correlated with the expression of *Let-7*, and it was shown that HMGA2 expression was high during early embryogenesis. During cancer progression when *Let-7* expression was lost, HMGA2 was upregulated, and experiments on a subgroup of cells of the NCI60 panel showed that this mechanism renders cells more mobile and invasive. Furthermore, cells seemed to be more resistant to the toxic effects of death ligands such as CD95L and Taxanes (Park et al., 2007). In a recent study, Trang et al. (2011) showed that synthetic miRNA mimics in complex with a novel neutral lipid emulsion can target lung tumors and demonstrated a therapeutic benefit in mouse models. Mice treated with the *Let-7* mimic showed a significant reduction in tumor mass (Trang et al., 2011; Trang et al., 2009). In nasopharyngeal carcinoma cells, the levels of

*Let-7*a, *Let-7*b, *Let-7*d, *Let-7*e, *Let-7*g, and *Let-7*i, were reduced in comparison to normal epithelial cells. All of these *Let-7* family members showed an anti-proliferative effect on nasopharyngeal carcinoma cell lines. It was shown that *Let-7*g was most effective in reducing the proliferative activity of these cells. One of the reasons therefore could be a downregulation of c-MYC. (Wong et al., 2011)

2) Impact of *Let-7* on Chemotherapy and Radiotherapy

Of most interest from a clinical point of view, there is increasing evidence that *Let-7* directly impacts on chemotherapy and radiotherapy response. For example, Meng et al. (2011) demonstrated that conserved *Let-7* and miR-181 family members were upregulated in hepatocellular cancer stem cells. An inhibition of *Let-7* increased the chemosensitivity of hepatocellular cancer stem cells to sorafenib and doxorubicin. Another group investigated the effect of *Let-7* on resistance to sorafenib-induced apoptosis. Sorafenib is a novel anti-cancer drug for heptocellular carcinoma (HCC), which is able to downregulate MCI1 expression that was reported to be overexpressed in HCC. Sorafenib treatment clearly downregulated MCL1 expression in hepatoma cells but did not affect MCL1 expression in normal hepatocytes. Of importance is the finding that sorafenib-induced apoptosis was markedly enhanced in *Let-7*c transfected cells (Shimizu et al., 2010). Moreover, *in vitro*, *Let-7* was shown to sensitize ovarian cancer cells to platinum-based chemotherapy, but to induce ovarian cancer resistance to paclitaxel (Lu et al., 2007). In another study, Tsang et al. (2008) identified Caspase-3 as the target of *Let-7*a in human hepatocellular carcinoma cells (HepG2) and squamous carcinoma cells (A431). Caspase 3 is the key executioner caspase in apoptosis. The authors examined the drug-induced apoptosis in HepG2 cells and A431 cells treated with doxorubicin, paclitaxel and interferon-gamma. Enforced *Let-7*a expression increased the resistance in A431 cells and HepG2 cells to apoptosis. To verify if the resistance to drug-induced apoptosis by *Let-7*a was mediated via caspase 3, the effect of caspase 3 inhibitors was examined. Overexpression of *Let-7* family members including *Let-7*a and down-regulation of caspase-3 were demonstrated in this context. Knockdown of *Let-7*a levels led to a rise of the caspase-3 levels. Furthermore, *Let-7*a knockdown also enhanced the doxorubicin-induced apoptosis in the cell lines, strengthening the hypothesis that *Let-7*a, through its effect on casapse-3, plays a role in modifying drug response of human cancer cells (Tsang et al., 2008). Similar findings regarding the effect of knocking-down *Let-7*a expression were described in another study. Inhibition of *Let-7*a increased in human malignant cholangiocytes chemotherapy-induced apoptosis in response to gemcitabine (Meng et al., 2007). Yu et al. (2011) investigated the effect of *Let-7* modulation on chemotherapy in head and neck cancer. By separating aldehyde dehydrogenase 1-positive cells (ALDH1^{+}) from ALDH1^{-} cells, the authors could demonstrate that *Let-7*a expression was reduced in ALDH1^{+} cells, whereas ALDH1^{-} cells showed higher expression. In addition, ALDH1^{+} cells were shown to be more resistant to cisplatin treatment, and expression levels of the drug-resistance genes ABCB1, ABCG2 and ABCG5 were higher in ALDH1^{+} cells than in ALDH1^{-} cells. Further experiments proved that overexpression of *Let-7* or knockdown of NANOG, an oral cancer stem expression marker in ALDH1^{+} cells, effectively blocked tumor metastasis and significantly prolonged survival time in ALDH1^{+}-transplanted immunocompromised mice (Yu et al., 2011). An alternate mechanism possibly contributing to the impact of *Let-7* on drug resistance might be regulation of IMP1. As mentioned above, *Let-7* targets this gene. Ovarian cancer patients presenting a relapse of disease showed a downregulation of *Let-7*, and an upregulation of IMP1 and MDR1, which resulted in an adverse prognosis of patients treated with Taxanes. The reduction of MDR1 increased sensitivity to the two MDR1 substrates Taxol and Vinblastine but not to Carboplatin (Boyerinas et al., 2011). Another receptor, which is highly expressed in several cancers, and known to be involved in chemoresistance is the progesterone receptor membrane component 1 (PGRMC1). A recently published study investigated PGRMC1 expression in ovarian cancer cells. It was shown that *Let-7*i targeted PGRMC1 in SKOV-3 cells. Progesterone had an effect on PGRMC1 expression by stimulation of *Let-7*i (Wendler et al., 2011). Also in breast cancer, there is first evidence that *Let-7* impacts on chemosensitivity. Estradiol promotes cell movement and invasion in this tumor type (Zheng, 2011). In MCF-7 breast cancer cells, Estradiol increased the expression of *Let-7*a, *Let-7*b, *Let-7*c, *Let-7*d, *Let-7*e, *Let-7*f, *Let-7*g, and *Let-7*i. Estradiol is a regulator of gene expression and a ligand for estrogen receptors. The authors concluded that Estradiol seems to be involved in regulation of *Let-7* expression. Moreover, they pointed out the tumor suppressor role of Estradiol might be mediated via regulation of miRNA expression, and it might be possible that Estradiol acts via induction of expression of *Let-7* and consequent reduction of expression of RAS and c-MYC (Bhat-Nakshatri et al., 2009). Oh et al. (2010) investigated the direct involvement of *Let-7* on radiation response. Overexpression of *Let-7*a led to a downregulation of K-RAS expression and increased radiosensitivity of A549 lung cancer cells. Furthermore, LIN28 (as described above) was found to be a posttranscriptional repressor of *Let-7* biogenesis. Inhibition of LIN28 increased the expression of *Let-7* and radiosensitized lung carcinoma (A549) and pancreatic cancer (ASPC1) cells (Oh et al., 2010).

Conclusion

MiRNAs are an incredible class of small RNA molecules, which exert great impact on global gene expression. *Let-7* is one of the most extensively studied miRNAs, and there is increasing evidence that this family of miRNAs enormously affects tumor initiation, growth and progression in a variety of malignancies. Based on these findings, there is now a growing interest in using these molecules for clinical purposes. In this book chapter, we demonstrated that several members of the *Let-7* family might play a crucial role in cancer detection, staging, planning of personalized therapy and, even more important, in the future treatment of a number of malignant diseases. We showed that *Let-7* correlates with prognosis and outcome in various cancers, and that resistant tumors (*in vitro* and *in vivo*) seem to present distinct *Let-7* expression patterns, making *Let-7* miRNAs promising tools for predicting outcome of cancer patients and response to anticancer treatment on an individual basis. Moreover, chemotherapy and irradiation obviously affects *Let-7* expression, which could be used to monitor treatment success during these therapies without time-wasting and expensive radiological examinations.

Most importantly, we highlighted first studies that investigated the impact of *Let-7* on tumor cell growth and on response to conventional treatment options. In this context, modulating *Let-7* expression might qualify as a new and promising first-line treatment option in cancer, or as an additive to chemotherapeutic or radiotherapeutic treatments, helping to augment the effects of these treatments and to overcome resistance which is a major obstacle in modern anticancer treatment.

REFERENCES

Akao, Y., Nakagawa, Y., Naoe, T. *Let-7* miRNA functions as a potential growth suppressor in human colon cancer cells. *Biol. Pharm. Bull.* 2006;29:903-906.

Ali, S., Almhanna, K., Chen, W., Philip, P. A., Sarkar, F. H. Differentially expressed miRNAs in the plasma may provide a molecular signature for aggressive pancreatic cancer. *Am. J. Transl. Res.* 2010;3:28-47.

Ambros, V. MiRNA pathways in flies and worms: growth, death, fat, stress, and timing. *Cell* 2003;113:673-676.

Bartel, D. P. MiRNAs: genomics, biogenesis, mechanism, and function. *Cell* 2004;116:281-297.

Bhat-Nakshatri, P., Wang, G., Collins, N. R., Thomson, M. J., Geistlinger, T. R., Carroll, J. S., Brown, M., Hammond, S., Srour, E. F., Liu, Y., Nakshatri, H. Estradiol-regulated miRNAs control estradiol response in breast cancer cells. *Nucleic Acids Res.* 2009;37: 4850-4861.

Boyerinas, B., Park, S. M., Murmann, A. E., Gwin, K., Montag, A. G., Zillardt, M. R., Hua, Y. J., Lengyel, E., Peter, M. E. *Let-7* modulates acquired resistance of ovarian cancer to Taxanes via IMP-1-mediated stabilization of MDR1. *J. Cancer.* 2011.

Brueckner, B., Stresemann, C., Kuner, R. et al. The human *Let-7*a-3 locus contains an epigenetically regulated miRNA gene with oncogenic function. *Cancer Res.* 2007;67: 1419-1423.

Chaudhry, M. A., Sachdeva, H., Omaruddin, R. A. Radiation-induced micro-RNA modulation in glioblastoma cells differing in DNA-repair pathways. *DNA Cell Biol.* 2010;29:553-561

Chaudhry, M. A. Biomarkers for human radiation exposure. *J. Biomed. Sci.* 2008;15:557-563..

Childs, G., Fazzari, M., Kung, G., Kawachi, N., Brandwein-Gensler, M., McLemore, M., Chen, Q., Burk, R. D., Smith, R. V., Prystowsky, M. B., Belbin, T. J., Schlecht, N. F. Low-level expression of miRNAs *Let-7*d and miR-205 are prognostic markers of head and neck squamous cell carcinoma. *Am. J. Pathol.* 2009 ;174:736-745.

Chin, L. J., Ratner, E., Leng, S. et al. A SNP in a *Let-7* miRNA complementary site in the KRAS 3' untranslated region increases non-small cell lung cancer risk. *Cancer Res.* 2008;68:8535-8540.

Christensen, B. C., Moyer, B. J., Avissar, M. et al. A *Let-7* miRNA-binding site polymorphism in the KRAS 3' UTR is associated with reduced survival in oral cancers. *Carcinogenesis* 2009;30:1003-1007.

Esquela-Kerscher, A., Trang, P., Wiggins, J. F., Patrawala, L., Cheng, A., Ford, L., Weidhaas, J. B., Brown, D., Bader, A. G., Slack, F. J. The *Let-7* miRNA reduces tumor growth in mouse models of lung cancer. *Cell Cycle.* 2008;7:759-764.

Garzon, R., Fabbri, M., Cimmino, A., Calin, G. A., Croce, C. M. MiRNA expression and function in cancer. *Trends Mol. Med.* 2006;12:580-587.

Graziano, F., Canestrari, E., Loupakis, F., Ruzzo, A., Galluccio, N., Santini, D., Rocchi, M., Vincenzi, B., Salvatore, L., Cremolini, C., Spoto, C., Catalano, V., D'Emidio, S., Giordani, P., Tonini, G., Falcone, A., Magnani, M. Genetic modulation of the *Let-7* miRNA binding to KRAS 3'-untranslated region and survival of metastatic colorectal cancer patients treated with salvage cetuximab-irinotecan. *Pharmacogenomics J.* 2010;10:458-464.

Griffiths-Jones, S., Saini, H. K., Van Dongen, S., Enright, A. J. MiRBase: tools for miRNA genomics. *Nucleic Acids Res.* 2008;36:D154-158.

Guan, H., Zhang, P., Liu, C., Zhang, J., Huang, Z., Chen, W., Chen, Z., Ni, N., Liu, Q., Jiang, A. Characterization and functional analysis of the human miRNA *Let-7*a2 promoter in lung cancer A549 cell lines. *Mol. Biol. Rep.* 2011;38:5327-5334.

Hummel, R., Hussey, D. J., Haier, J. MiRNAs: predictors and modifiers of chemo- and radiotherapy in different tumour types. *Eur. J. Cancer.* 2010;46:298-311.

Johnson, S. M., Grosshans, H., Shingara, J. et al. RAS is regulated by the *Let-7* miRNA family. *Cell* 2005;120:635-647.

L. Lu, D. Katsaros, I. De la Longrais, O. Sochirca and H. Yu, Hypermethylation of *Let-7*a-3 in epithelial ovarian cancer is associated with low insulin-like growth factor-II expression and favorable prognosis. *Cancer Res.* 2007;67:10117-10122.

Li, Y., VandenBoom, T. G. 2nd, Kong, D., Wang, Z., Ali, S., Philip, P. A., Sarkar, F. H. Upregulation of miR-200 and *Let-7* by natural agents leads to the reversal of epithelial-to-mesenchymal transition in gemcitabine-resistant pancreatic cancer cells. *Cancer Res.* 2009;69:6704-6712.

Lu, L., Katsaros, D., Shaverdashvili, K., Qian, B., Wu, Y., De la Longrais, I. A., Preti, M., Menato, G., Yu, H. Pluripotent factor LIN28 and its homologue LIN28b in epithelial ovarian cancer and their associations with disease outcomes and expression of *Let-7*a and IGF-II. *Eur. J. Cancer.* 2009;45:2212-2218.

Lu, L., Katsaros, D., Wiley, A. et al. Promoter-specific transcription of insulin-like growth factor-II in epithelial ovarian cancer. *Gynecol. Oncol.* 2006; 103: 990-995.

Lu, L., Schwartz, P., Scarampi, L., Rutherford, T., Canuto, E. M., Yu, H., Katsaros, D. MiRNA *Let-7*a: A potential marker for selection of paclitaxel in ovarian cancer management. *Gynecol. Oncol.* 2011;122:366-371.

Meng, F., Glaser, S. S., Francis, H., Demorrow, S., Han, Y., Passarini, J. D., Stokes, A., Cleary, J. P., Liu, X., Venter, J., Kumar, P., Priester, S., Hubble, L., Stoloch, D., Sharma, J., Liu, C. G., Alpini, G. Functional Analysis of miRNAs in Human Hepatocellular Cancer Stem Cells. *J. Cell Mol. Med.* 2011;1582-4934.

Meng, F., Henson, R., Wehbe-Janek, H., Smith, H., Ueno, Y., Patel, T. The MiRNA *Let-7*a modulates interleukin-6-dependent STAT-3 survival signaling in malignant human cholangiocytes. *J. Biol. Chem.* 2007;282:8256-8264.

Mulero-Navarro, S., Esteller, M. Epigenetic biomarkers for human cancer: the time is now. *Crit. Rev. Oncol. Hematol.* 2008;68:1–11.

Oh, J. S., Kim, J. J., Byun, J. Y., Kim, I. A. LIN28-let7 modulates radiosensitivity of human cancer cells with activation of K-Ras. *Int. J. Radiat. Oncol. Biol. Phys.* 2010;76:5-8.

Park, S. M., Shell, S., Radjabi, A. R., Schickel, R., Feig, C., Boyerinas, B., Dinulescu, D. M., Lengyel, E., Peter, M. E. *Let-7* prevents early cancer progression by suppressing expression of the embryonic gene HMGA2. *Cell Cycle.* 2007;6:2585-2590.

Pramanik, D., Campbell, N. R., Karikari, C., Chivukula, R., Kent, O. A., Mendell, J. T., Maitra, A. Restitution of tumor suppressor miRNAs using a systemic nanovector inhibits pancreatic cancer growth in mice. *Mol. Cancer Ther.*_2011;10:1470-1480.

Qing, L., Guo-dong, L., Xu, Q., Yue-hua, G., Shu-tao, Z., Tao, L., Xiao-mei, L. Role of miRNA *Let-7* and effect to HMGA2 in esophageal squamous cell carcinoma. *Mol. Biol. Rep.* 2011.

Raponi, M., Dossey, L., Jatkoe, T., Wu, X., Chen, G., Fan, H., Beer, D. G. MiRNA classifiers for predicting prognosis of squamous cell lung cancer. *Cancer Res.* 2009;69:5776-5783.

Reinhart, B. J. et al. The 21-nucleotide *Let-7* RNA regulates developmental timing in Caenorhabditis elegans. *Nature* 2000; 403: 901-906.

Roush, S., Slack, F. J. The *Let-7* family of miRNAs. *Trends Cell Biol.* 2008; 18(10):505-516.

Sampson, V. B., Rong, N. H., Han, J., Yang, Q., Aris, V., Soteropoulos, P., Petrelli, N. J., Dunn, S. P., Krueger, L. J. MiRNA *Let-7*a down-regulates MYC and reverts MYC-induced growth in Burkitt lymphoma cells. *Cancer Res.* 2007;67:9762-9770.

Shell, S., Park, S. M., Radjabi, A. R., Schickel, R., Kistner, E. O., Jewell, D. A. et al. *Let-7* expression defines two differentiation stages of cancer. *Proc. Natl. Acad. Sci.* U 2007;104:11400-11405.

Shimizu, S., Takehara, T., Hikita, H., Kodama, T., Miyagi, T., Hosui, A., Tatsumi, T., Ishida, H., Noda, T., Nagano, H., Doki, Y., Mori, M., Hayashi, N. The *Let-7* family of miRNAs inhibits Bcl-xL expression and potentiates sorafenib-induced apoptosis in human hepatocellular carcinoma. *J. Hepatol.* 2010;52:698-704.

Takamizawa, J., Konishi, H., Yanagisawa, K., Tomida, S., Osada, H., Endoh, H. et al. Reduced expression of the *Let-7* miRNAs in human lung cancers in associationwith shortened postoperative survival. *Cancer Res.* 2004;64:3753-3756.

Takamizawa, J. K. H., Yanagisawa, K. et al. Reduced expression of the *Let-7* miRNAs in human lung cancers in association with shortened postoperative survival. *Cancer Res.* 2004; 64: 3753-3756.

Torrisani, J., Bournet, B., Du Rieu, M. C., Bouisson, M., Souque, A., Escourrou, J., Buscail, L., Cordelier, P. *Let-7* MiRNA transfer in pancreatic cancer-derived cells inhibits in vitro cell proliferation but fails to alter tumor progression. *Hum. Gene. Ther.* 2009;20:831-844.

Trang, P., Medina, P. P., Wiggins, J. F., Ruffino, L., Kelnar, K., Omotola, M., Homer, R., Brown, D., Bader, A. G., Weidhaas, J. B., Slack, F. J. Regression of murine lung tumors by the *Let-7* miRNA. *Oncogene.* 2010;29:1580-1587.

Trang, P., Wiggins, J. F., Daige, C. L., Cho, C., Omotola, M., Brown, D., Weidhaas, J. B., Bader, A. G., Slack, F. J. Systemic delivery of tumor suppressor miRNA mimics using a neutral lipid emulsion inhibits lung tumors in mice. *Mol. Ther.* 2011;19:1116-1122.

Tsang, W. P., Kwok, T. T. *Let-7*a miRNA suppresses therapeutics-induced cancer cell death by targeting caspase-3. *Apoptosis.* 2008;13:1215-1222.

Tsujiura, M., Ichikawa, D., Komatsu, S., Shiozaki, A., Takeshita, H., Kosuga, T., Konishi, H., Morimura, R., Deguchi, K., Fujiwara, H., Okamoto, K., Otsuji, E. Circulating miRNAs in plasma of patients with gastric cancers. *Br. J. Cancer* 2010;102:1174-1179.

Weidhaas, J. B., Babar, I., Nallur, S. M., Trang, P., Roush, S., Boehm, M., Gillespie, E., Slack, F. J. MiRNAs as potential agents to alter resistance to cytotoxic anticancer therapy. *Cancer Res.* 2007;67:11111-11116.

Wendler, A., Keller, D., Albrecht, C., Peluso, J. J., Wehling, M. Involvement of *Let-7*/miR-98 miRNAs in the regulation of progesterone receptor membrane component 1 expression in ovarian cancer cells. *Oncol. Rep.* 2011;25:273-279.

Wong, T. S., Man, O. Y., Tsang, C. M., Tsao, S. W., Tsang, R. K., Chan, J. Y., Ho, W. K., Wei, W. I., To, V. S. MiRNA *Let-7* suppresses nasopharyngeal carcinoma cells proliferation through downregulating c-Myc expression. *J. Cancer Res. Clin. Oncol.* 2011;137:415-422.

Yanaihara, N., Caplen, N., Bowman, E. et al. Unique miRNA molecular profiles in lung cancer diagnosis and prognosis. *Cancer Cell* 2006;9:189-198.

Yang, N., Kaur, S., Volinia, S., Greshock, J., Lassus, H., Hasegawa, K. et al. MiRNA microarray identifies *Let-7*i as a novel biomarker and therapeutic target in human epithelial ovarian cancer. *Cancer Res.* 2008;68:10307-10314.

Yu, C. C., Chen, Y. W., Chiou, G. Y., Tsai, L. L., Huang, P. I., Chang, C. Y., Tseng, L. M., Chiou, S. H., Yen, S. H., Chou, M. Y., Chu, P. Y., Lo, W. L. MiRNA *Let-7*a represses chemoresistance and tumourigenicity in head and neck cancer via stem-like properties ablation. *Oral Oncol.* 2011;47:202-210.

Zhang, B., Pan, X., Cobb, G. P., Anderson, T. A. miRNAs as oncogenes and tumor suppressors. *Dev. Biol.* 2007;302:1–12.

Zhang, B., Pan, X. RDX induces aberrant expression of miRNAs in mouse brain and liver. *Environ. Health Perspect.* 2009;117:231-240.

Zhang, W., Winder, T., Ning, Y., Pohl, A., Yang, D., Kahn, M., Lurje, G., Labonte, M. J., Wilson, P. M., Gordon, M. A., Hu-Lieskovan, S., Mauro, D. J., Langer, C., Rowinsky, E. K., Lenz, H. J. A *Let-7* miRNA-binding site polymorphism in 3'-untranslated region of KRAS gene predicts response in wild-type KRAS patients with metastatic colorectal cancer treated with cetuximab monotherapy. *Ann. Oncol.* 2011;22:104-109.

Zheng, S., Huang, J., Zhou, K., Zhang, C., Xiang, Q., Tan, Z., Wang, T., Fu, X. 17β-Estradiol Enhances Breast Cancer Cell Motility and Invasion via Extra-Nuclear Activation of Actin-Binding Protein Ezrin. *PLoS One.* 2011;6:e22439.

In: MicroRNA *Let-7*
Editor: Neetu Dahiya

ISBN: 978-1-62081-152-8

Chapter 7

THERAPEUTIC POTENTIAL OF MICRORNA *LET-7* IN HUMAN DISEASES

Elaine Lu Wang,[1,2] Xu-Hui Li[3] and Zhi Rong Qian[1,*]

[1]Department of Human Pathology, Institute of Health Biosciences, The University of Tokushima Graduate School. 3-18-15 Kuramoto-cho, Tokushima, Japan

[2]Department of Legal Medicine, Kanazawa Medical University, 1-1 Daigaku, Uchinada, Ishikawa, Japan

[3]Protein Science Laboratory of the Ministry of Education, School of Life Sciences, Tsinghua University, Beijing, China

1. ABSTRACT

MicroRNAs (MiRNAs) are emerging as important therapeutic molecules. Their role in regulating fundamental processes of life and altered expression in a number of human diseases has made it necessary to study their diagnostic and therapeutic potential. A number of human cancers shows unique miRNA signature useful in distinguishing different subtypes of tumors. MiRNAs has been reported to function both as oncogenes and tumor suppressors. For example members of *Let-7* family are well known tumor suppressors whereas miR-17-92 family members function as oncogenes. Most of the *Let-7* family members are downregulated in cancers and two well know oncogenic targets regulated by *Let-7* are RAS and HMGA2. *Let-7* is also involved in development of drug resistance. In addition to cancer altered expression of *Let-7* has been reported in other diseases such as lung fibrosis, inflammation etc. Because of its involvement in development and progression of the diseases, *Let-7* is being studied for its potential as a therapeutic target. This chapter summarizes the current knowledge of *Let-7* functions in cancer and other disease conditions and challenges associated with development of miRNA based therapeutic approaches.

* Corresponding author: Zhi Rong Qian. Department of Human Pathology, Institute of Health Biosciences, The University of Tokushima Graduate School, 3-18-15 Kuramoto-cho, Tokushima 770-8503, Japan. Email: zrqian@basic.med.tokushima-u.ac.jp.

2. Introduction to MiRNAs

1) MiRNAs

MiRNAs are a relatively recently identified non-coding RNAs. Mature miRNAs are short RNA molecules, approximately 19-23 nucleotides in length. MiRNAs are excised from 60-110 nucleotide RNA precursor molecules, and then undergo successive enzymatic processing by Drosha, Pasha, and Dicer into the mature form. MiRNAs can regulate the expression of a large number of genes at post-transcriptional level through sequence-specific binding to the 3' untranslated regions (UTRs) of target mRNAs, resulting in the inhibition and/or degradation of target mRNAs (Bartel, 2004; Filipowicz et al., 2008). miRNAs have been identified in a wide range of species, and computational analysis shows that nearly 30% of protein-coding genes can be modulated by miRNAs (Filipowicz et al., 2008). In general, miRNAs negatively regulate the expression of their targets, however, it is also reported that miR-369-3p can up-regulate the expression of its target, tumor necrosis factor-α (TNF-α) (Bartel, 2005). miRNAs have been demonstrated to play an important role in many biological processes, such as cell cycle control, proliferation, apoptosis, differentiation, metabolism, hemopoiesis, and development (Ambros, 2004).

2) Specific MiRNAs as Oncogenes/Tumor Suppressors

A rapidly growing body of evidence shows that miRNAs have comprehensive functions in tumor progression. Some miRNAs may function as oncogenes (also called oncomiRNAs) while some miRNAs are supposed to be tumor suppressors (Ambros, 2004). The importance of miRNAs in cancer is highlighted by the fact that half of all miRNA genes are located in cancer-associated regions or fragile sites, which are frequently altered or deleted in cancer (Calin et al., 2004). Many tumor types show unique miRNA signatures, thus miRNAs may be useful in cancer diagnosis and even distinguishing tumor subtypes (Calin et al., 2004; Nelson and Weiss, 2008). MiRNAs contribute to oncogenesis by functioning as Tumor Suppressor Genes or as oncogenes directly or indirectly. Because miRNAs are negative regulators of gene expression, the changes of the expression level of these miRNAs can be tumorigenic if they target mRNAs from a tumor suppressor gene, as well as an oncogene (Vasudevan et al., 2007; Gartel and Kendel, 2008).

I) MiRNAs and Apoptosis

One of the central pathways that prevent tumorigenesis is the p53 tumor suppressor protein, which induces growth arrest, mediates DNA repair and promotes apoptosis. Conversely, mutations in the p53 pathway are found in nearly all cancers. Recently, miRNA joined the p53 regulatory network after the miR-34 family members were shown to be the direct transcriptional targets of p53 (He et al., 2007). Surprisingly, miR-34 activation recapitulates elements of p53 activity, including induction of cell cycle arrest and the promotion of apoptosis, whereas loss of miR-34 impairs p53-mediated cell death (He et al., 2007). In addition, p53 itself is upregulated by miR-29 family members, which directly suppress phosphatidylinositol 3-kinase p85a and GTPase CDC42, two negative regulators of p53 (Park et al., 2009). Conversely, p53 is suppressed by miR-125b (Lee et al., 2009).

II) MiRNAs Are Involved in Invasion, Metastasis and Tumor Angiogenesis

Invasion and metastasis are responsible for >90% of cancer-related mortality. Epithelial-mesenchymal transition (EMT), mainly characterized by E-cadherin degradation, is an important event in invasion and metastasis. E-cadherin, which is encoded by E-cadherin gene (CDH1), is a well-known molecule witch maintains the cell-to-cell junctions in epithelial cells (Cano et al., 2000; Comijn et al., 2001). Its aberrant expression has been implicated in cancer progression and metastasis (Bremnes et al., 2002; Masterson and O'Dea, 2007). MiR-373 was considered to have an effective role in E-cadherin targeting. It has been confirmed that the transfection of miR-373 and its precursor hairpin RNA into PC-3 cells readily induced E-cadherin expression. Together with miR-520c, they are considered to be metastasis-promoting miRNAs by direct suppression of CD44 which is consistently reduced in metastatic breast, colon and prostate cancers (Place et al., 2008; Huang et al., 2008). In breast cancer cells, it has been shown that miR-9 could directly target CDH1, the E-cadherin encoding mRNA, leading to increased cell motility and invasiveness (Ma et al., 2007).

MiR-21 was found up-regulated in many solid tumors. When anti-sense oligo against miR-21 was introduced into metastatic cancer cells, their metastatic ability was inhibited, as gauged by a tail-vein metastasis assay and a chick-embryo chorioallantoic-membrane metastasis assay. With further study, it was found that miR-21 could target phosphatase and tensin homolog (PTEN), tumor suppressor gene tropomyosin 1 (TPM1), and programmed cell death 4 (PDCD4), to promote the metastasis of cancers (Chan et al., 2005; Zhu et al., 2007; Asangani et al., 2008).

Meanwhile, some miRNAs play a role in inhibiting tumor invasion and metastasis. The miR-200 family, organized as two clusters in the genome, was expressed during EMT and able to hinder EMT by enhancing E-cadherin transcriptional expression through directly targeting ZEB1 and ZEB2 (Park et al., 2008; Burk et al., 2008). Martello have shown that high levels of miR-103/miR-107 were associated with metastasis and poor outcome in human breast cancer. At the cellular level, a key event fostered by miR-103/miR-107 was the induction of EMT, attained by down-regulating miR-200 levels (Korpal et al., 2008).

Angiogenesis frequently happens in various tumors and is highly related with invasion, metastasis and poor outcome (Kerbel 2008; Urbich et al., 2008). Early studies have indicated the contribution of specific miRNAs (e.g. miR-21, miR-155, and miR-126) to vascular diseases (Urbich et al., 2008). At present, miR-126 is widely accepted as an important factor for angiogenesis. Two groups published their investigations back-to-back that miR-126 could regulate the response of endothelial cells to vascular endothelial growth factor (VEGF), and regulate vascular integrity and angiogenesis in vivo (Fish et al., 2008). Nicoli et al. (2010) reported that zinc finger transcription factor KLF2A induced the expression of miR-126 leading to the activation of the VEGF signaling pathway.

III) MiRNAs and Cancer-Related Inflammation

Chronic inflammation is a major cause of cancer. The oncogenic mechanisms in chronic inflammation are complicated and not fully revealed. Some studies have suggested that induced epigenetic changes and genomic instability were involved in inflammation-induced carcinogenesis. miRNAs have emerged as a critical regulatory factor in the mammalian immune system. Genetic ablation of the miRNA machinery, as well as loss or deregulation of certain individual miRNAs, severely compromises immune response leading to immune disorders like autoimmunity and cancer (Xiao et al., 2009).

miR-146a is one of the first miRNAs identified to be involved in the regulation of immune function. Lipo- polysaccharide (LPS)-induced induction of miR-146a was observed in several other cell lines of myeloid origin, but not in B cell lines, suggesting the LPS-induced regulation of miR-146a is cell-type specific. In addition to the effect on Toll-like receptor (TLR) ligands, miR-146a was induced by TNF-α and interleukin (IL)-1β in an NF-κB dependent manner (Taganov et al., 2006; Perry et al., 2008). Many other miRNAs, such as miR-9, miR-21, miR-101, miR-155, miR-192, and miR-203, have been also associated with inflammatory diseases or immune responses (Sonkoly and Pivarcsi, 2009; Schetter et al., 2010).

IV) MiRNAs Are Involved in Drug Resistance

Drug resistance remains a major clinical obstacle to successful treatment. The understanding of the drug resistance mechanism is important for the cancer therapy. Recently, the role of miRNA in drug resistance and drug sensitivity was reported: a polymorphism in the miR-24 mRNA binding site, 829C →T in dihydrofolate reductase (DHFR) 3' UTR, led to loss of miR-24 function and resulted in DHFR overexpression and methotrexate resistance (Mishra et al., 2007). Blower et al studied the effects on chemotherapy sensitivity of 3 miRNAs (*Let-7*i, miR-16, and miR-21) that were previously implicated in cancer biology (Blower et al., 2008). Significant overexpression of 8 miRNAs and down-regulation of 7 miRNAs were detected in a tamoxifen-resistant breast cancer cell line com- pared with the tamoxifen-sensitive cell line (Miller et al., 2008). Kovalchuk et al. (2008) reported that miR-451 regulated the expression of multidrug resistance 1 (MDR-1) gene in the doxorubicin-resistant human breast adenocarcinoma cell line MCF-7. In ovarian cancer, Yang et al. (2008) demonstrated that the up-regulation of miR-214 promoted the survival of ovarian cancer cells and induced resistance of cisplatin. Sorrentino et al. (2008) showed that a panel of miRNAs (*Let-7*e, miR-30c, -125b, -130a and -335) were diversely expressed in all the resistant cell lines of ovarian cancer. In non-small cell lung cancer, 5 miRNAs (miR-15b, miR-100, miR-125b, miR-221, and miR-222) were found to up-regulate in the resistant cell lines (Garofalo et al., 2008). Bertino et al. (2007) put forward the concept of miRNA pharmacogenomics. This model can be defined as the study miRNAs and the miRNAsNPs/polymorphisms in their target genes may determine drug behavior in order to improve efficiency of drugs. Upon reaching a deeper understanding of the mechanism of miRNA in anticancer drug resistance, miRNAs might well fulfill their promise as valuable therapeutics in overcoming anticancer drug resistance.

3) MiRNAs in other Disease

Obesity is the result of a chronic imbalance between energy intake and expenditure. This leads to storage of excess energy in adipocytes, typically exhibiting both hypertrophy and hyperplasia. Adipocyte hypertrophy, evident in both overweight patients and patients with type 2 diabetes, was originally considered to be the sole route whereby adipose tissue mass increased in adults. However, adipocyte hyperplasia (adipogenesis) is now known to contribute to the increased adipose tissue mass in obesity (de Ferranti S and Mozaffarian, 2008). Recent studies reported that miRNAs regulate adipocyte differentiation, oxidative stress in adipose tissues of obese persons. The miR-17-92 cluster, miR-21, miR-103, miR-143, miR-371, and miR-378/miR-378* increase adipogenesis, characterized by the up-regulation of adipogenic markers and by increased triglycerides (Bork et al., 2010; Gerin et al., 2010; Kim et al., 2009; Wang et al., 2008). In contrast, *Let-7*, miR-27, miR-130, miR-138, miR-369-5p, and miR-448 inhibit adipogenic differentiation, evidenced by the down-regulation of adipogenic factors and by a decrease in triglycerides (KInoshita et al., 2010; Lee et al., 2011; Lin et al., 2009; Sun et al., 2009). Impairment of adipocyte function is associated with endoplasmic reticulum and mitochondrial oxidative stress, which further aggravates adipose tissue dysfunction. Dysfunctional adipocytes also exhibit an inflammatory phenotype, with increased production of proinflammatory adipocytokines and decreased production of anti-inflammatory adipocytokines.

Initially, miR-375 was identified among several miRNAs expressed selectively in pancreatic endocrine cell lines (Poy et al., 2004). In addition, studies ex vivo revealed that overexpression of miR-375 resulted in suppressed glucose-stimulated insulin secretion, whereas inhibition of miR-375 enhanced insulin secretion, both through the suppression by miR-375 of myotrophin and pyruvate dehydrogenase kinase, isozyme 1 (PDK1) (Poy et al., 2004; El Ouaamari et al., 2008). Mice lacking miR-375 are hyperglycaemic: they exhibit increased total numbers of pancreatic a-cells, fasting and fed plasma glucagon levels and increased gluconeogenesis and hepatic glucose output, whereas pancreatic b-cell mass is decreased as a result of impaired proliferation (Poy et al., 2009).

Several studies have investigated mice with conditional knockout of Dicer in various post-mitotic neuronal cell types, including Prukinje cells, midbrain dopaminergic neurons and cortical and hippocampal neurons. These studies all demonstrated a progressive loss of neurons (Davis et al., 2008; Kim et al., 2007). In these studies, miR-133b was specifically identified as a miRNA expressed in midbrain dopaminergic neurons, but deficient in midbrain tissue from patients with Parkinson's disease (Davis et al., 2008; Kim et al., 2007). It was suggested that miR-133b regulated the maturation and function of dopaminergic neurons within a negative feedback circuit via suppression of paired-like homeodomain 3 (Pitx3) (Davis et al., 2008; Kim et al., 2007. The first evidence describing a role for the miRNA machinery in preventing systemic autoimmunity used a strain of mice homozygous for a hypomorphic variant of the roquin gene (sanroque mice) that develop high titres of autoantibodies and a lupus-like systemic autoimmunity, as well as autoimmune diabetes against a genetically susceptible background (Vinuesa et al., 2005). Roquin is a newly discovered component involved in miRNA-mediated regulation. Hypomorphic mutant Roquin in sanroque mice led to impaired miRNA activity. Such impairment resulted in decreased miR-101 /103-mediated mRNA decay of the inducible T cell costimulator (Icos); the higher Icos level promoted lymphoproliferation (Yu et al., 2007). Mice subjected to T cell-specific Dicer or Drosha deletion developed colitis and lung and liver inflammation from 4 months of age, with a > 50% reduction in numbers of regulatory T (Treg) cells. These cells are a specialized subpopulation of T cells that act to suppress activation of the immune system and thereby maintain immune system homeostasis and tolerance to self-antigens (Chong et al., 2008; Cobb et al., 2006).

3. INTRODUCTION *LET-7*

1) The Discovery of *Let-7* and its Role in Development

Although *Let-7* was found as the second miRNA after lin-4 in *Caenorhabditis elegans* (Reinhart et al., 2000), its high conservation across the animal phylogeny from *C. elegans* to human provided the clue for the generality of miRNAs as essential regulators of gene expression in various organisms (Pasquinelli et al., 2000). To date, there are a total of 9169 mature miRNAs found across 103 species, of which 885 miRNAs are found in humans (miRBase Release 13.0 http://miRNA.sanger.ac.uk/sequences). Many of these miRNAs are highly conserved across different species. A good example is the *Let-7* family of miRNAs, which is highly conserved across diverse animal species from worms to humans (Pasquinelli et al., 2000). Consistent with its role in regulating cell proliferation and differentiation during development in different species, the deregulation of this miR has been shown as a feature of many types of cancer, reflecting the major conserved roles of *Let-7*. *Let-7* was initially identified as a heterochronic gene by forward genetics in *C. elegans* (Reinhart et al., 2000). During *C. elegans* development, the hypodermal skin cells known as seam cells undergo asymmetrical division at each larval stage.

Recent studies in *Drosophila* have shown that *Let-7* also functions as a heterochronic gene in this species (Caygill and Johnston, 2008; Sokol et al., 2008). In *Drosophila*, there is only a single *Let-7* gene (Lagos-Quintana et al., 2001), and it becomes expressed at the end of the third larval instar stage and peaks in pupae during metamorphogenesis (Pasquinelli et al., 2000) Like *C. elegans*, *Drosophila* undergo a series of molting processes in their development, and a pulse of ecdysone is released before each molting stage.

In human, these different members are *Let-7*a-1, *Let-7*a-2, *Let-7*a-3, *Let-7*b, *Let-7*c, *Let-7*d, *Let-7*e, let-f7-1, *Let-7*f-2, *Let-7*g, *Let-7*i, miR-98, and miR-202 (Ruby et al., 2006). Among the members, *Let-7*a has identical sequence across various animal species from *C. elegans* to human. The increase in *Let-7* expression in late developmental stages has been reported in many organisms (Sempere et al., 2002; Lancman et al., 2005; Liu et al., 2007; Wulczyn et al., 2007). Previously, studies have shown the expression patterns of *Let-7* in vertebrates during development; however, the direct contribution of *Let-7* in development has not been demonstrated (Lancman et al., 2005; Liu et al., 2007; Wulczyn et al., 2007; Schulman et al., 2005). This is probably because vertebrate *Let-7* family members are likely to have redundant. In mammals, *Let-7* levels increase during embryogenesis and during brain development (Wulczyn et al., 2007; Schulman et al., 2005). *Let-7* is undetectable in human and mouse embryonic stem cells, and the level of *Let-7* increases upon differentiation (Wulczyn et al., 2007; Thomson et al., 2004). This high expression of *Let-7* is then maintained in various adult tissues (Thomson et al., 2004; Sempere et al., 2004). Conversely, the reduction of *Let-7* levels has been found in many human cancers, which is reflective of the reverse embryogenesis process that occurs during tumorigenesis (Park et al., 2007). As we will discuss in the following sections, while *Let-7* was initially viewed as one single activity, emerging data suggest that the *Let-7* family contains miRNAs with different activities. Therefore, we will name the specific *Let-7* family member (wherever possible) throughout this review.

2) *Let-7* in Cancer

Let-7 family members are located at chromosomal regions that are often altered or deleted in human tumors (Calin et al., 2004). Down-regulation of *Let-7* expression has been reported in breast, lung, colon and others cancers (Iorio et al., 2005; Takamizawa et al., 2004; Akao et al., 2006; Shell et al., 2007), and *Let-7* is regarded as a tumor suppressor by targeting RAS oncogene and HMGA2 (Johnson et al., 2005; Mayr et al., 2007). *Let-7* is widely viewed as a tumor suppressor miRNA. Consistent with this activity, the expression of *Let-7* family members is downregulated in many cancer types when compared to normal tissue and during tumor progression.

The loss of *Let-7* family members also has prognostic value as it indicates poor survival. A general downregulation of *Let-7* was found to correlate with poor survival in lung cancer (Takamizawa et al., 2004). Specifically, this was also reported for *Let-7*a-2 (Yanaihara et al., 2006). Low expression of *Let-7*d (combined with low miR-205 expression) was found in head and neck squamous cell carcinoma (HNSCC) patients, and was predictive of poor survival (Childs et al., 2009), and a combined loss of *Let-7*d with an increase in expression of the *Let-7* target high mobility group 2A (HMGA2) was indicative of poor survival in ovarian cancer (Shell et al., 2007). Although less frequent, upregulation of certain *Let-7* family members has also been observed, suggesting that *Let-7* does not play a tumor suppressor function under all circumstances and/or in all tissues.

The upregulation of *Let-7*b and *Let-7*i was associated with high grade transformation in lymphoma (Lawrie et al., 2009), indicating that increased expression of *Let-7* family members could be used as a prognostic marker to identify patients at risk of high grade transformation, or for higher grade cancer. Hypomethylation of the *Let-7*a-3 locus was found to cause higher expression of *Let-7*a-3 in epithelial ovarian cancer (Lu et al., 2007) and lung cancer (Brueckner et al., 2007). Hypomethylation did not only cause increased expression of *Let-7*a-3, but subsequently deregulated the expression of other genes, including oncogenes and genes involved in cell proliferation, adhesion, and differentiation (Brueckner et al., 2007).

The conflicting data on the deregulation of *Let-7* in various cancers indicate that the function of the *Let-7* family is not clearly defined, or that individual *Let-7* family members can have different activities. An example of this is a report on malignant mesothelioma, where *Let-7*b* was found to be highly expressed but *Let-7*e* was severely reduced (Guled et al., 2009). Cases such as this suggest that *Let-7* family members are potentially subject to differential regulation within the same cell, and an important area the miRNA field will need to address is whether different family members indeed have specific activities in a particular cell type, or whether tissue-specific regulation is the most important mechanism utilized to obtain specific *Let-7*-mediated cellular outcomes regardless of which *Let-7* member is involved.

3) Cancer-Relevant *Let-7* Targets

I) RAS

The miRNAs that are encoded by the *Let-7* family were the first group of oncomiRNAs shown to regulate the expression of an oncogene, specifically the RAS genes. RAS proteins are membrane-associated GTPase signaling proteins that regulate cellular growth and differentiation. About 15–30% of human tumors possess mutations in RAS genes, and activating mutations that result in the increased expression of RAS cause cellular transformation. Therefore, a miRNA that regulates the expression level of these potentially oncogenic proteins is predicted to keep the rate of cellular proliferation in check.

The first mammalian target of *Let-7* was identified by virtue of evolutionary sequence conservation between the nematode *C. elegans* and humans. Using a computational screen for *C. elegans* 3'-UTR sequences containing *Let-7* family complementary sites (LCS), Johnson et al. (2005) identified let-60 as a top-scoring candidate whose human ortholog is the RAS oncogene. The three different RAS oncogenes are frequently deregulated in many human cancers. It was determined that let-60 expression is tightly and directly regulated by miR-84 in *C. elegans* vulval cells, and it was also demonstrated that human *Let-7* does specifically target RAS in human cancer cells. The same group determined that *Let-7* family members *Let-7*a, *Let-7*c, and *Let-7*g are significantly decreased in lung cancer tumors, and that many *Let-7* family members are located in genomic regions frequently deleted in lung cancer patients. This evidence, together with the fact that RAS is frequently up-regulated in human lung tumor samples, suggested a pivotal role for the *Let-7* family in the suppression of oncogenic RAS proteins in vivo. This direct targeting of RAS by *Let-7* was confirmed in non-small cell lung cancer (NSCLC), where it was demonstrated in a mouse model that *Let-7*g inhibited tumor growth via suppression of RAS (Kumar et al., 2008).

II) HMGA2

HMGA2 is often subject to a chromosomal rearrangement that results in loss of the C-terminal region, which was thought to be the cause of cellular transformation (Schoenmakers et al., 1995). Rearrangement also causes separation of the open reading frame and the 3'-UTR of the gene transcript, which suggests that neoplastic evolution, may be due to loss of repressive elements at the 3'-UTR (Borrmann et al., 2001). HMGA2 takes part in many diverse biological processes such as embryogenesis, differentiation, and neoplastic transformation (Sgarra et al., 2004). Overexpression of HMGA2 is a hallmark of various tumors and is associated with highly malignancy (Fusco et al., 2007). Recently, some studies revealed that high-mobility group A2 (HMGA2) is negatively regulated by the *Let-7* miRNAs family in vitro (Mayr et al., 2007; Lee and Dutta, 2007).

There are seven *Let-7* target sites within the 3'-UTR of HMGA2, indicating that *Let-7* may play a role in its regulation (Shell et al., 2007). Four recent studies demonstrated the direct regulation of HMGA2 by *Let-7*, which was mediated by interaction at the 3'-UTR in various cell types such as lung carcinoma (H1299), retinoblastoma (Y79)(Lee and Dutta, 2007), cervical cancer (HeLa) (Mayr et al., 2007), liver cancer (HepG2), ovarian cancer (Shell et al., 2007), uterine leiomyoma and prostate cancer (Peng et al., 2008; Dangi-Garimella et al., 2009). An inverse relationship between *Let-7* and HMGA2 expression was illustrated *in vivo* in uterine leiomyomas, with overexpression of HMGA2 and low *Let-7* expression common to large leiomyomas (Peng et al., 2008). Furthermore, regulation of HMGA2 by *Let-7* reduced cell proliferation in prostate cancer cells (Dangi-Garimella et al., 2009), while disruption of this regulation increased anchorage-independent growth and caused tumor formation in mice (Mayr et al., 2007). This suggests that *Let-7* inhibits cellular proliferation by negatively regulating HMGA2, disruption of this regulation may be an important event in promotion of tumor growth. Intriguingly, an inverse correlation between

HMGA2 and *Let-7* was confirmed in pituitary adenoma and neuroendocrine tumor of digestive tract. HMGA2 overexpression and the decrease of *Let-7* were significantly correlated with tumor proliferation, growth, invasion and tumor grade (Qian et al., 2009; Rahman et al., 2009).

4) *Let-7* in Other Disease

I) Let-7 in Lung Fibrosis

Because transforming growth factor b1 (TGFb1) is a key regulator of lung fibrosis (Sheppard, 2006) we scanned the promoters of differentially expressed miRNAs for SMAD binding elements (SBE) and found a SBE upstream of the miRNA *Let-7*d. We confirmed physical binding of SMAD3 to the putative *Let-7*d promoter by chromatin immunoprecipitation using an antibody to SMAD3 and SMAD3 electrophoretic mobility shift assay. *Let-7*d localized to alveolar epithelial cells of normal lungs, and the number of *Let-7*d positive cells positively correlated with the forced vital capacity, suggesting that *Let-7*d is required for normal lung function. Inhibition of *Let-7*d in vitro and in vivo by intratracheal administration of an antagomiR that inhibits all members of the *Let-7* family resulted in upregulation of the mesenchymal markers and downregulation of epithelial markers suggesting EMT. Interestingly, HMGA2, a known target of *Let-7*d (Mayr et al., 2007) and a regulator of EMT (Thuault et al., 2006; Thuault et al., 2008) was induced when *Let-7*d was inhibited *in vitro* and is highly expressed in alveolar epithelial cells of patients with IPF but not in controls. The molecular mechanism by which *Let-7* inhibition may lead to fibrosis is still unclear. Suppression of KRAS, MYC, cyclin D2, CDK6, and CDC25A along with HMGA2 may be equally important in the antifibrotic effect of *Let-7*. Although we have demonstrated that inhibition of *Let-7* miRNAs leads to a gain of a profibrotic phenotype in lung epithelial cells *in vitro* and changes consistent with early fibrotic changes *in vivo*, we have not yet shown that administration of *Let-7*d or its mimic is sufficient to prevent or reverse fibrosis.

II) MiRNAs Regulating Oxidative Stress and Inflammation in Relation to Obesity and Atherosclerosis

Approximately 400 studies have demonstrated changes in miRNA expression profiles in relation to cardiovascular diseases (Fichtlscherer et al., 2010; Matkovich et al.,2009; Ceolotto et al., 2011), of which a few have identified miRNAs that may be useful to protect against the development of these diseases (Skog et al., 2008). We can conclude that *Let-7*, miR-17, miR-21, miR-27b, miR-34a, miR-92, miR-125b, miR-130a, miR-132, miR-143, miR-155, miR-210, miR-221, and miR-222 are deregulated in both adipose and vascular tissues (Hulsmans et al., 2011). Furthermore, Sun et al. (2009) reported that *Let-7* inhibits adipogenesis by inhibiting HMGA2.

III) Let-7 in Cancer-related Inflammation

Iliopoulos et al. (2009) found that *Let-7* directly inhibited IL6 expression, resulting in higher levels of IL6 than achieved by NF-κB activation in several cancer cell lines. This result suggests that *Let-7* might be important to the cancer-related inflammation.

5) *Let-7* and Drug Sensitivity

Chemotherapy is one of the most frequently utilized treatment modalities for various forms of human cancer. Unfortunately, the majority of patients in most forms of cancer relapse within 5 years, and recurrent disease is frequently much more resistant to treatment via chemotherapeutic agents. Understanding the mechanisms via which drug resistance evolves in treatment-refractory cancers is critically important in the fight to reduce cancer-related mortalities. Changes in miRNA expression profiles, due to their profound effect on gene expression signatures, are emerging as an intriguing mechanism for the development of chemoresistance in many cancers. A group of miRNAs has been implicated in modulation of survival pathways and/or apoptotic response in cancer cells. As described above, the *Let-7* family of miRNAs plays a role in a host of cellular functions, and that includes modulation of drug sensitivity. The most direct mechanistic link between a *Let-7* family member and drug sensitivity involves *Let-7*a, which has been shown to directly target caspase-3. *Let-7*a, which is overexpressed in some human cancers, induces resistance to a variety of drugs that induce caspase-3-dependent apoptosis (Tsang et al., 2008). Interestingly, four independent studies have found a correlation between *Let-7*i expression and either sensitivity or resistance to certain compounds (Blower et al., 2008; Yang et al., 2008).

Let-7 family members have also been shown to be involved in radiation response of cancer cell lines. When A549 lung cancer cells were exposed to ionizing radiation, the expression of all *Let-7* family members decreased except for *Let-7*g, whose expression increased in response to radiation. (Weidhaas et al., 2007). Furthermore, another study showed more than four fold induction of *Let-7*f and *Let-7*i levels Jurkat T-cell leukemia cells in response to ionizing radiation (Chaudhry, 2009). Recent evidence has demonstrated that breast cancer T-ICs can be characterized by low *Let-7* expression, and that modulating *Let-7* expression in these cells alters their stem-like properties (Yu et al., 2007). This phenomenon was also observed in comma-Db, an immortalized but not transformed mouse mammary epithelial cell line that contains a permanent population of undifferentiated progenitor cells that are able to repopulate the mouse mammary tree. These progenitor cells were found to be *Let-7* low, and enforced *Let-7* expression eliminated the self-renewing cells from the population (Ibarra et al., 2007). Taken together, these studies suggest that a *Let-7* low status is common to both T-ICs and normal epithelial progenitor cells.

6) LIN28 Regulates *Let-7* Expression in Cancer

LIN28 is a specific, posttranscriptional inhibitor of *Let-7* biogenesis. Nam et al. (2011) report crystal structures of mouse LIN28 in complex with sequences from *Let-7*d, *Let-7*-f1, and *Let-7*g precursors. The two folded domains of LIN28 recognize two distinct regions of the RNA and are sufficient for inhibition of *Let-7 in vivo*. They also show by NMR spectroscopy that the linker connecting the two folded domains is flexible, accommodating LIN28 binding to diverse *Let-7* family members. Protein-RNA complex formation imposes specific conformations on both components that could affect downstream recognition by other processing factors.

Furthermore, Piskounova et al. (2011) found that LIN28A and LIN28B selectively block the expression of *Let-7* miRNAs and function as oncogenes in a variety of human cancers.

LIN28A recruits a TUTase (Zcchc11/TUT4) to *Let-7* precursors to block processing by Dicer in the cell cytoplasm. LIN28B functions in the nucleus by sequestering primary *Let-7* transcripts and inhibiting their processing by the Microprocessor. The inhibitory effects of Zcchc11 depletion on the tumorigenic capacity and metastatic potential of human cancer cells and xenografts are restricted to LIN28A-expressing tumors. Furthermore, the majority of human colon and breast tumors analyzed exclusively express either LIN28A or LIN28B. LIN28A is expressed in HER2-overexpressing breast tumors, whereas LIN28B expression characterizes triple-negative breast tumors.

6. MiRNAs in Disease Therapy

According to the recent findings on miRNAs, RNA inhibition can be used to treat cancer patients in 2 ways: (1) by mimicking miRNAs using RNA molecules to directly target mRNA of genes involved in cancer pathogenesis, and (2) by directly targeting miRNAs that participate in cancer pathogenesis. Most RNA inhibition agents used in recent preclinical and clinical studies have been antisense oligonucleotides (ASOs), ribozymes and DNAzymes, small-interfering RNAs (siRNAs) and short hairpin RNAs (shRNAs), miRNAs and anti-miRNA agents (eg, ASO anti-miRNAs), locked nucleic acid (LNA) anti-miRNAs, or antagomiRNAs (Gleave and Monia, 2005; Bumcrot et al., 2006; Oplinska and Gewirtz, 2002; Doudna and Lorsch, 2005).

1) *Let-7* and other MiRNAs as Therapeutic Agents

I) MiRNAs as New Therapeutic Agents in MiRNA Mimic Therapy

Decreased expression of a miRNA subgroup is present in certain cancers where loss of a miRNA inhibitory effect contributes to oncogene activation. In these situations, a rational treatment is to inhibit expression of the constitutively overexpressed tumor proteins using synthetic miRNA mimics. This approach was shown to work *in vitro*. In SKOV3 ovarian cancer cells, synthesized miRNA-14U was able to inhibit the overexpressed HER2 protein (Tsuda et al., 2005). The disadvantages of this approach are that miRNA-mimic oligonucleotides have only a transient effect, are unstable, and may require repeated supplementation. A different technique that bypasses the shortcomings of miR-mimic oligos uses vector-based miRNA expression to produce stably expressed miRNAs. This approach has been used to increase *Let-7* expression, which is significantly reduced in lung cancer cell lines and lung cancer tissues. Takamizawa et al. (2004) designed expression constructs to synthesize the mature miRNAs of two predominant *Let-7* isoforms, *Let-7*a and *Let-7*f, under the control of the RNA polymerase-III H1-RNA gene promoter. When the *Let-7* expression vector was introduced into the A549 adenocarcinoma cell line, a 78.6%reduction in the number of colonies was observed. In two separate experiments, an overexpression vector bearing *Let-7* in lung cancer cell lines enhanced lung cancer cell radiosensitivity (Weidhaas et al., 2007), altered cell cycle progression, and reduced cell division (Johnson et al., 2007).

II) Let-7 as Inhibitor in Cancer Therapy

Lung carcinomas represent the leading cause of cancer deaths in human worldwide; therefore, it might represent a target for miRNA mimic agents. One study found that the *Let-7* family of miRNAs is highly downregulated in lung cancers (Yanaihara et al., 2006) and that reduced *Let-7* expression is significantly correlated with shorter survival times after potentially curative resection in lung cancer (Takamizawa et al., 2004). Furthermore, because *Let-7* negatively regulates RAS and because overexpression of RAS-p21 protein is associated with shorter survival (Johnson et al., 2005), administering *Let-7* agents to patients with lung cancers could be a new therapeutic approach. Supporting this view, intranasal *Let-7* administration reduced tumor formation *in vivo* in the lungs of animals that express a G12D activating mutation for the K-ras oncogene (Esquela-Kerscher et al., 2008). To date, there have been no reports of miRNA mimic agents being used in clinical practice to reproduce the effects of endogenous miRNAs and include all the chemical modifications necessary for stability and processing. However, we envision that restoring the expression of miRNAs that are downregulated in cancer will be possible in the clinic.

III) Delivery Issues in Using MiRNAs as Therapeutic Agents

There are two major practical issues in the delivery of RNAs as therapeutic molecules (Spizzo et al., 2009). The first issue is the best formula for these new agents: should naked RNAs or a vehicle be used for delivery, and if a vehicle is used, should it be a viral or nonviral vehicle? Concerning naked RNA, two options have been tested: (1) the use of siRNA duplexes 21 nucleotides long that have 2-nucleotide 3' overhangs, and (2) the use of longer siRNAs of 27 mers and shRNAs (29 nucleotides). A second major practical issue in the delivery of RNA is the route of administration. Local administration has certain advantages, such as the need for lower RNA doses, less severe adverse or toxic reactions, and better bioavailability of the drug to the target tissue. Recently, numerous publications have described the local delivery of siRNA molecules to tumors in vivo, including direct administration of siRNA in the peritoneal cavity, testes, or transurethral (Aigner et al., 2006), for some cancers, such as leukemias, local administration cannot be achieved, making systemic delivery mandatory.

2) *Let-7* and other MiRNAs as Therapeutic Targets

I) Various Agents Used for RNA Inhibition

Anti-miRNA oligonucleotides (AMOs) are ASOs that target a specific miRNA. AMOs are RNA molecules that have a reverse complementary sequence to the miRNAs that they target. The binding of AMOs to miRNAs impairs the interaction between miRNA and target mRNAs. ASO inhibition has been found to be a powerful technique to inhibit miRNAs (Weiler et al., 2006).

Locked nucleic acids comprise a new class of bicyclic high-affinity RNAs mimicking a N-type (C3'-endo) conformation by the introduction of a 2'-O4-C methylene bridge. Several studies have demonstrated that LNA-modified oligonucleotides exhibited high thermal stability, when they hybridize with their RNA target molecules (Kumar et al., 1998). Importantly, LNA incorporation generally improves mismatch discrimination compared with unmodified reference oligonucleotides, and LNAs mediate high-affinity hybridization (using the Watson- Crick rules) without compromising base pairing selectivity. One study showed that miR-21 was strongly overexpressed in glioblastomas (Ciafre et al., 2005), and another study showed that miR-21 could be silenced in vitro using LNA-modified AMOs, which leads to a significant reduction in cell viability accompanied by elevated intracellular caspase levels (Chan et al., 2005). LNA antimRNA oligonucleotides have also been tested *in vivo* in mouse models and been proven to be safe and effective.

An efficient *in vivo* miRNA loss-of-function pharmacologic approach was recently developed by producing a novel class of chemically engineered oligonucleotides termed "antagomiRNAs," which act as specific and effective silencers of miRNA expression in mice (Krutzfeldt et al., 2005). Single-stranded 23-nucleotide RNA molecules complementary to the targeted miR-122 were modified to increase the stability of the RNA and protect it from degradation. Furthermore, *in vivo* miR-122 inhibition by antagomiRNAs induced a significant decrease in cholesterol plasma levels, in hepatic fatty acid and cholesterol synthesis rates, whereas, antagomiR-122 used in a diet-induced obesity mouse model resulted in decreased plasma cholesterol levels and a reduction in several lipogenic genes.

Some small molecules have been proven to work against proteins involved in cancers. Recently, an assay for small-molecule inhibitors of miRNA function was developed, and potential inhibitors of the miRNA pathway were found (GumiReddy et al., 2008). For instance, a hit compound structurally related to diazobenzene 1 was found to reduce the expression of miR-21 in HeLa cells and did not display cytotoxic effects at a concentration of 10 mmol.

II) Let-7 as Target in Cancer and other Disease Therapy

Recent study reported that *Let-7* family can control glucose homeostasis and insulin sensitivity (Frost and Olson, 2011). Diabetes mellitus is the most common metabolic disorder worldwide and a major risk factor for cardiovascular disease. MiRNAs are negative regulators of gene expression that have been implicated in many biological processes, including metabolism. Recent findings demonstrate that *Let-7* regulates multiple aspects of glucose metabolism and suggest antimiR-induced *Let-7* knockdown as a potential treatment for type 2 diabetes mellitus. Global and pancreas-specific overexpression of *Let-7* in mice resulted in impaired glucose tolerance and reduced glucose-induced pancreatic insulin secretion. Mice overexpressing *Let-7* also had decreased fat mass and body weight, as well as reduced body size. Global knockdown of the *Let-7* family with an antimiR was sufficient to prevent and treat impaired glucose tolerance in mice with diet-induced obesity. AntimiR treatment of mice on a high-fat diet also resulted in increased lean and muscle mass, but not increased fat mass, and prevented ectopic fat deposition in the liver.

Conclusion

MiRNAs were linked to the progression of all types of human tumors and other certain diseases that were investigated to date. These discoveries could be exploited for the development of useful markers for diagnosis and prognosis, as well as for the development of new RNA-based cancer therapies. Based on the link between abnormal miRNA expression and diseases, two major miRNA-based therapeutic strategies are: (i) to restore the expression of the miRNA reduced in diseases; and (ii) conversely, blunting overexpressed miRNA. As our understanding of miRNA continues to evolve, prospects for miRNA-based therapy will improve.

As we have discussed throughout this review, the *Let-7* family of miRNAs plays a role in an exceedingly diverse array of cellular activities. There is a very clear link between *Let-7* expression and poorly differentiated, aggressive cancers and other diseases. As we have elucidated, *Let-7* expression is regulated on multiple levels, and particular family members appear to be specifically deregulated in certain cancers or up-regulated in other certain diseases. In addition, while the evidence is growing that loss of *Let-7* increases resistance to certain chemotherapeutic drugs and to radiation. Recent available data suggest that restoration of *Let-7* expression to tumors where it has been lost holds great therapeutic potential for the treatment of these aggressive types of cancer. Although exciting, the use of miRNAs/anti-miRNAs in human cancer therapy has yet to show high efficiency in the inhibition of target RNAs, significant increase in patient survival, and low toxic effects.

References

Aigner, A. Gene silencing through RNA interference (RNAi) in vivo: strategies based on the direct application of siRNAs. *J. Biotechnol.* 2006;124:12-25.

Akao, Y., Nakagawa, Y., Naoe, T. *Let-7* microRNA functions as a potential growth suppressor in human colon cancer cells. *Biol. Pharm. Bull.* 2006;29:903-906.

Ambros, V. The functions of animal microRNAs. *Nature*. 2004;431:350-355.

Asangani, I. A., Rasheed, S. A., Nikolova, D. A. et al. MicroRNA-21 (miR-21) post-transcriptionally downregulates tumor suppressor Pdcd4 and stimulates invasion, intravasation and metastasis in colorectal cancer. *Oncogene*. 2008;27:2128-2136.

Bartel, B. MicroRNAs directing siRNA biogenesis. *Nat. Struct. Mol. Biol.* 2005;12:569-571.

Bartel, D. P. MicroRNAs: genomics, biogenesis, mechanism, and function. *Cell.* 2004;116:281-297.

Bertino, J. R., Banerjee, D., Mishra, P. J. Pharmacogenomics of microRNA: a miRNAsNP towards individualized therapy. *Pharmacogenomics*. 2007;8:1625-1627.

Blower, P. E., Chung, J. H., Verducci, J. S. et al. MicroRNAs modulate the chemosensitivity of tumor cells. *Mol. Cancer Ther.* 2008;7:1-9.

Bork, S., Horn, P., Castoldi, M. et al. Adipogenic differentiation of human mesenchymal stromal cells is down-regulated by microRNA-369-5p and up-regulated by microRNA-371. *J. Cell Physiol.* 2011;226:2226-2234.

Borrmann, L., Wilkening, S., Bullerdiek, J. The expression of HMGA genes is regulated by their 3'UTR. *Oncogene*. 2001;20:4537-4541.

Bremnes, R. M., Veve, R., Hirsch, F. R. et al. The E-cadherin cell-cell adhesion complex and lung cancer invasion, metastasis, and prognosis. *Lung Cancer*. 2002;36:115-124.

Brueckner, B., Stresemann, C., Kuner, R. et al. The human *Let-7*a-3 locus contains an epigenetically regulated microRNA gene with oncogenic function. *Cancer Res.* 2007;67:1419-1423.

Bumcrot, D., Manoharan, M., Koteliansky, V. et al. RNAi therapeutics: a potential new class of pharmaceutical drugs. *Nat. Chem. Biol.* 2006;2:711-719.

Burk, U., Schubert, J., Wellner, U. et al. A reciprocal repression between ZEB1 and members of the miR-200 family promotes EMT and invasion in cancer cells. *EMBO Rep.* 2008;9:582-589.

Calin, G. A., Sevignani, C., Dumitru, C. D. et al. Human microRNA genes are frequently located at fragile sites and genomic regions involved in cancers. *Proc. Natl. Acad. Sci.* U S A. 2004;101:2999-3004.

Cano, A., Perez-Moreno, M. A., Rodrigo, I. et al. The transcription factor snail controls epithelial-mesenchymal transitions by repressing E-cadherin expression. *Nat. Cell Biol.* 2000;2:76-83.

Caygill, E. E., Johnston, L. A. Temporal regulation of metamorphic processes in *Drosophila* by the *Let-7* and miR-125 heterochronic microRNAs. *Curr. Biol.* 2008;18:943-950.

Ceolotto, G., Papparella, I., Bortoluzzi, A. et al. Interplay between miR-155, AT1R A1166C polymorphism, and AT1R expression in young untreated hypertensives. *Am. J. Hypertens.* 2011;24:241-246.

Chan, J. A., Krichevsky, A. M., Kosik, K. S. MicroRNA-21 is an antiapoptotic factor in human glioblastoma cells. *Cancer Res.* 2005;65:6029-6033.

Chaudhry, M. A. Real-time PCR analysis of micro-RNA expression in ionizing radiation-treated cells. *Cancer Biotherand Radio Pharma.* 2009;24:49-56.

Childs, G., Fazzari, M., Kung, G. et al. Low-level expression of microRNAs *Let-7*d and miR-205 are prognostic markers of head and neck squamous cell carcinoma. *Am. J. Pathol.* 2009;174:736-745.

Chong, M. M., Rasmussen, J. P., Rudensky, A. Y. et al. The RNAseIII enzyme Drosha is critical in T cells for preventing lethal inflammatory disease. *J. Exp. Med.* 2008;205:2005-2017.

Ciafre, S. A., Galardi, S., Mangiola, A. et al. Extensive modulation of a set of microRNAs in primary glioblastoma. *Biochem. Biophys. Res. Commun.* 2005;334:1351-1358.

Cobb, B. S., Hertweck, A., Smith, J. et al. A role for Dicer in immune regulation. *J. Exp. Med.* 2006;203:2519-2527.

Comijn, J., Berx, G., Vermassen, P. et al. The two-handed E box binding zinc finger protein SIP1 downregulates E-cadherin and induces invasion. *Mol. Cell.* 2001;7:1267-1278.

Dangi-Garimella, S., Yun, J., Eves, E. M. et al. Raf kinase inhibitory protein suppresses a metastasis signalling cascade involving LIN28 and *Let-7*. *EMBO J.* 2009;28:347-358.

Davis, T. H., Cuellar, T. L., Koch, S. M. et al. Conditional loss of Dicer disrupts cellular and tissue morphogenesis in the cortex and hippocampus. *J. Neurosci.* 2008;28:4322-4330.

De Ferranti, S., Mozaffarian, D. The perfect storm: obesity, adipocyte dysfunction, and metabolic consequences. *Clin. Chem.* 2008;54:945-955.

Doudna, J. A., Lorsch, J. R. Ribozyme catalysis: not different, just worse. *Nat. Struct. Mol. Biol.* 2005;12:395-402.

El Ouaamari, A., Baroukh, N., Martens, G. A. et al. miR-375 targets 3'-phosphoinositide-dependent protein kinase-1 and regulates glucose-induced biological responses in pancreatic beta-cells. *Diabetes.* 2008;57:2708-2717.

Esquela-Kerscher, A., Trang, P., Wiggins, J. F. et al. The *Let-7* microRNA reduces tumor growth in mouse models of lung cancer. *Cell Cycle.* 2008;7:759-764.

Fichtlscherer, S., De Rosa, S., Fox, H. et al. Circulating microRNAs in patients with coronary artery disease. *Circ. Res.* 2010;107:677-684.

Filipowicz, W., Bhattacharyya, S. N., Sonenberg, N. Mechanisms of post-transcriptional regulation by microRNAs: are the answers in sight? *Nat. Rev. Genet.* 2008;9:102-114.

Fish, J. E., Santoro, M. M., Morton, S. U. et al. miR-126 regulates angiogenic signaling and vascular integrity. *Dev. Cell.* 2008;15:272-284.

Frost, R. J., Olson, E. N. Control of glucose homeostasis and insulin sensitivity by the *Let-7* family of microRNAs. *Proc. Natl. Acad. Sci.* U S A. 2011;108:21075-21080.

Fusco, A., Fedele, M. Roles of HMGA proteins in cancer. *Nat. Rev. Cancer.* 2007;7:899-910.

Garofalo, M., Quintavalle, C., Di Leva, G. et al. MicroRNA signatures of TRAIL resistance in human non-small cell lung cancer. *Oncogene.* 2008;27:3845-3855.

Gartel, A. L., Kandel, E. S. miRNAs: Little known mediators of oncogenesis. *Semin. Cancer Biol.* 2008;18:103-110.

Gerin, I., Bommer, G. T., McCoin, C. S. et al. Roles for miRNA-378/378* in adipocyte gene expression and lipogenesis. *Am. J. Physiol. Endocrinol. Metab.* 2010;299:E198-206.

Gleave, M. E., Monia, B. P. Antisense therapy for cancer. *Nat. Rev. Cancer.* 2005;5:468-479.

Guled, M., Lahti, L., Lindholm, P. M. et al. CDKN2A, NF2, and JUN are dysregulated among other genes by miRNAs in malignant mesothelioma -A miRNA microarray analysis. Genes *Chromosomes Cancer.* 2009;48:615-623.

GumiReddy, K., Young, D. D., Xiong, X. et al. Small-molecule inhibitors of microrna miR-21 function. *Angew. Chem. Int. Ed. Engl.* 2008;47:7482-7484.

He, L., He, X., Lowe, S. W. et al. microRNAs join the p53 network--another piece in the tumor-suppression puzzle. *Nat. Rev. Cancer.* 2007;7:819-822.

Huang, Q., GumiReddy, K., Schrier, M. et al. The microRNAs miR-373 and miR-520c promote tumor invasion and metastasis. *Nat. Cell Biol.* 2008;10:202-210.

Hulsmans, M., De Keyzer, D., Holvoet, P. MicroRNAs regulating oxidative stress and inflammation in relation to obesity and atherosclerosis. *FASEB J.* 2011;25:2515-2527.

Ibarra, I., Erlich, Y., Muthuswamy, S. K. et al. A role for microRNAs in maintenance of mouse mammary epithelial progenitor cells. *Genes Dev.* 2007;21:3238-3243.

Iliopoulos, D., Hirsch, H. A., Struhl, K. An epigenetic switch involving NF-kappaB, LIN28, *Let-7* MicroRNA, and IL6 links inflammation to cell transformation. *Cell.* 2009;139:693-706.

Iorio, M. V., Ferracin, M., Liu, C. G. et al. MicroRNA gene expression deregulation in human breast cancer. *Cancer Res.* 2005;65:7065-7070.

Johnson, C. D., Esquela-Kerscher, A., Stefani, G. et al. The *Let-7* microRNA represses cell proliferation pathways in human cells. *Cancer Res.* 2007;67:7713-7722.

Johnson, S. M., Grosshans, H., Shingara, J. et al. RAS is regulated by the *Let-7* microRNA family. *Cell.* 2005;120:635-647.

Kerbel, R. S. Tumor angiogenesis. *N Engl. J. Med.* 2008;358:2039-2049.

Kim, J., Inoue, K., Ishii, J. et al. A MicroRNA feedback circuit in midbrain dopamine neurons. *Science.* 2007;317:1220-1224.

Kim, Y. J., Hwang, S. J., Bae, Y. C. et al. MiR-21 regulates adipogenic differentiation through the modulation of TGF-beta signaling in mesenchymal stem cells derived from human adipose tissue. *Stem. Cells.* 2009;27:3093-3102.

Kinoshita, M., Ono, K., Horie, T. et al. Regulation of adipocyte differentiation by activation of serotonin (5-HT) receptors 5-HT2AR and 5-HT2CR and involvement of microRNA-448-mediated repression of KLF5. *Mol. Endocrinol.* 2010;24:1978-1987.

Korpal, M., Lee, E. S., Hu, G. et al. The miR-200 family inhibits epithelial-mesenchymal transition and cancer cell migration by direct targeting of E-cadherin transcriptional repressors ZEB1 and ZEB2. *J. Biol. Chem.* 2008;283:14910-14914.

Kovalchuk, O., Filkowski, J., Meservy, J. et al. Involvement of microRNA-451 in resistance of the MCF-7 breast cancer cells to chemotherapeutic drug doxorubicin. *Mol. Cancer Ther.* 2008;7:2152-2159.

Krutzfeldt, J., Rajewsky, N., Braich, R. et al. Silencing of microRNAs in vivo with antagomiRNAs'. *Nature.* 2005;438:685-689.

Kumar, M. S., Erkeland, S. J., Pester, R. E. et al. Suppression of non-small cell lung tumor development by the *Let-7* microRNA family. *Proc. Natl. Acad. Sci.* U S A. 2008;105:3903-3908.

Kumar, R., Singh, S. K., Koshkin, A. A. et al. The first analogues of LNA (locked nucleic acids): phosphorothioate-LNA and 2'-thio-LNA. *Bioorg. Med. Chem. Lett.* 1998;8:2219-2222.

Lagos-Quintana, M., Rauhut, R., Lendeckel, W. et al. Identification of novel genes coding for small expressed RNAs. *Science.* 2001;294:853-858.

Lancman, J. J., Caruccio, N. C., Harfe, B. D. et al. Analysis of the regulation of lin-41 during chick and mouse limb development. *Dev. Dyn.* 2005;234:948-960.

Lawrie, C. H., Chi, J., Taylor, S. et al. Expression of microRNAs in diffuse large B cell lymphoma is associated with immunophenotype, survival and transformation from follicular lymphoma. *J. Cell Mol. Med.* 2009;13:1248-1260.

Le, M. T., Teh, C., Shyh-Chang, N. et al. MicroRNA-125b is a novel negative regulator of p53. *Genes Dev.* 2009;23:862-876.

Lee, E. K., Lee, M. J., Abdelmohsen, K. et al. miR-130 suppresses adipogenesis by inhibiting peroxisome proliferator-activated receptor gamma expression. *Mol. Cell Biol.* 2011;31:626-638.

Lee, Y. S., Dutta, A. The tumor suppressor microRNA *Let-7* represses the HMGA2 oncogene. *Genes Dev.* 2007;21:1025-1030.

Lin, Q., Gao, Z., Alarcon, R. M. et al. A role of miR-27 in the regulation of adipogenesis. *FEBS J.* 2009;276:2348-2358.

Liu, S., Xia, Q., Zhao, P. et al. Characterization and expression patterns of *Let-7* microRNA in the silkworm (Bombyx mori). *BMC Dev. Biol.* 2007;7:88.

Lu, L., Katsaros, D., De la Longrais, I. A. et al. Hypermethylation of *Let-7*a-3 in epithelial ovarian cancer is associated with low insulin-like growth factor-II expression and favorable prognosis. *Cancer Res.* 2007;67:10117-10122.

Ma, L., Teruya-Feldstein, J., Weinberg, R. A. Tumour invasion and metastasis initiated by microRNA-10b in breast cancer. *Nature.* 2007;449:682-688.

Masterson, J., O'Dea, S. Posttranslational truncation of E-cadherin and significance for tumor progression. *Cells Tissues Organs.* 2007;185:175-179.

Matkovich, S. J., Van Booven, D. J., Youker, K. A. et al. Reciprocal regulation of myocardial microRNAs and messenger RNA in human cardiomyopathy and reversal of the microRNA signature by biomechanical support. *Circulation.* 2009;119:1263-1271.

Mayr, C., Hemann, M. T., Bartel, D. P. Disrupting the pairing between *Let-7* and HMGA2 enhances oncogenic transformation. *Science*. 2007;315:1576-1579.

Miller, T. E., Ghoshal, K., Ramaswamy, B. et al. MicroRNA-221/222 confers tamoxifen resistance in breast cancer by targeting p27Kip1. *J. Biol. Chem*. 2008;283:29897-29903.

Mishra, P. J., Humeniuk, R., Longo-Sorbello, G. S. et al. A miR-24 microRNA binding-site polymorphism in dihydrofolate reductase gene leads to methotrexate resistance. *Proc. Natl. Acad. Sci*. U S A. 2007;104:13513-13518.

Nam, Y., Chen, C., Gregory, R. I. et al. Molecular Basis for Interaction of *Let-7* MicroRNAs with LIN28. *Cell*. 2011;147:1080-1091.

Nelson, K. M., Weiss, G. J. MicroRNAs and cancer: past, present, and potential future. *Mol. Cancer Ther*. 2008;7:3655-3660.

Nicoli, S., Standley, C., Walker, P. et al. MicroRNA-mediated integration of haemodynamics and Vegf signalling during angiogenesis. *Nature*. 2010;464:1196-1200.

Opalinska, J. B., Gewirtz, A. M. Nucleic-acid therapeutics: basic principles and recent applications. *Nat. Rev. Drug Discov*. 2002;1:503-514.

Park, S. M., Gaur, A. B., Lengyel, E. et al. The miR-200 family determines the epithelial phenotype of cancer cells by targeting the E-cadherin repressors ZEB1 and ZEB2. *Genes Dev*. 2008;22:894-907.

Park, S. M., Shell, S., Radjabi, A. R. et al. *Let-7* prevents early cancer progression by suppressing expression of the embryonic gene HMGA2. *Cell Cycle*. 2007;6:2585-2590.

Park, S. Y., Lee, J. H., Ha, M. et al. miR-29 miRNAs activate p53 by targeting p85 alpha and CDC42. *Nat. Struct. Mol. Biol*. 2009;16:23-29.

Pasquinelli, A. E., Reinhart, B. J., Slack, F. et al. Conservation of the sequence and temporal expression of *Let-7* heterochronic regulatory RNA. *Nature*. 2000;408:86-89.

Peng, Y., Laser, J., Shi, G. et al. Antiproliferative effects by *Let-7* repression of high-mobility group A2 in uterine leiomyoma. *Mol. Cancer Res*. 2008;6:663-673.

Perry, M. M., Moschos, S. A., Williams, A. E. et al. Rapid changes in microRNA-146a expression negatively regulate the IL-1beta-induced inflammatory response in human lung alveolar epithelial cells. *J. Immunol*. 2008;180:5689-5698.

Piskounova, E., Polytarchou, C., Thornton, J. E. et al. LIN28A and LIN28B Inhibit *Let-7* MicroRNA Biogenesis by Distinct Mechanisms. *Cell*. 2011;147:1066-1079.

Place, R. F., Li, L. C., Pookot, D. et al. MicroRNA-373 induces expression of genes with complementary promoter sequences. *Proc. Natl. Acad. Sci*. U S A. 2008;105:1608-1613.

Poy, M. N., Eliasson, L., Krutzfeldt, J. et al. A pancreatic islet-specific microRNA regulates insulin secretion. *Nature*. 2004;432:226-230.

Poy, M. N., Hausser, J., Trajkovski, M. et al. miR-375 maintains normal pancreatic alpha- and beta-cell mass. *Proc. Natl. Acad. Sci*. U S A. 2009;106:5813-5818.

Qian, Z. R., Asa, S. L., Siomi, H. et al. Overexpression of HMGA2 relates to reduction of the *Let-7* and its relationship to clinicopathological features in pituitary adenomas. *Mod. Pathol*. 2009;22:431-441.

Rahman, M. M., Qian, Z. R., Wang, E. L. et al. Frequent overexpression of HMGA1 and 2 in gastroenteropancreatic neuroendocrine tumors and its relationship to *Let-7* downregulation. *Br. J. Cancer*. 2009;100:501-510.

Reinhart, B. J., Slack, F. J., Basson, M. et al. The 21-nucleotide *Let-7* RNA regulates developmental timing in Caenorhabditis elegans. *Nature*. 2000;403:901-906.

Ruby, J. G., Jan, C., Player, C. et al. Large-scale sequencing reveals 21U-RNAs and additional microRNAs and endogenous siRNAs in C. elegans. *Cell.* 2006;127:1193-1207.

Schetter, A. J., Heegaard, N. H., Harris, C. C. Inflammation and cancer: interweaving microRNA, free radical, cytokine and p53 pathways. *Carcinogenesis.* 2010;31:37-49.

Schoenmakers, E. F., Wanschura, S., Mols, R. et al. Recurrent rearrangements in the high mobility group protein gene, HMGI-C, in benign mesenchymal tumors. *Nat. Genet.* 1995;10:436-444.

Schulman, B. R., Esquela-Kerscher, A., Slack, F. J. Reciprocal expression of lin-41 and the microRNAs *Let-7* and miR-125 during mouse embryogenesis. *Dev. Dyn.* 2005;234:1046-1054.

Sempere, L. F., Dubrovsky, E. B., Dubrovskaya, V. A. et al. The expression of the *Let-7* small regulatory RNA is controlled by ecdysone during metamorphosis in Drosophila melanogaster. *Dev. Biol.* 2002;244:170-179.

Sempere, L. F., Freemantle, S., Pitha-Rowe, I. et al. Expression profiling of mammalian microRNAs uncovers a subset of brain-expressed microRNAs with possible roles in murine and human neuronal differentiation. *Genome Biol.* 2004;5:R13.

Sgarra, R., Rustighi, A., Tessari, M. A. et al. Nuclear phosphoproteins HMGA and their relationship with chromatin structure and cancer. *FEBS Lett.* 2004;574:1-8.

Shell, S., Park, S. M., Radjabi, A. R. et al. *Let-7* expression defines two differentiation stages of cancer. *Proc. Natl. Acad. Sci.* U S A. 2007;104:11400-11405.

Sheppard, D. Transforming growth factor beta: a central modulator of pulmonary and airway inflammation and fibrosis. *Proc. Am. Thorac. Soc.* 2006;3:413-417.

Skog, J., Wurdinger, T., Van Rijn, S. et al. Glioblastoma microvesicles transport RNA and proteins that promote tumor growth and provide diagnostic biomarkers. *Nat. Cell Biol.* 2008;10:1470-1476.

Sokol, N. S., Xu, P., Jan, Y. N. et al. *Drosophila Let-7* microRNA is required for remodeling of the neuromusculature during metamorphosis. *Genes Dev.* 2008;22:1591-1596.

Sonkoly, E., Pivarcsi, A. microRNAs in inflammation. *Int. Rev. Immunol.* 2009;28:535-561.

Sorrentino, A., Liu, C. G., Addario, A. et al. Role of microRNAs in drug-resistant ovarian cancer cells. *Gynecol. Oncol.* 2008;111:478-486.

Spizzo, R., Rushworth, D., Guerrero, M. et al. RNA inhibition, microRNAs, and new therapeutic agents for cancer treatment. *Clin. Lymphoma Myeloma.* 2009;9:S313-318.

Sun, T., Fu, M., Bookout, A. L. et al. MicroRNA *Let-7* regulates 3T3-L1 adipogenesis. *Mol. Endocrinol.* 2009;23:925-931.

Taganov, K. D., Boldin, M. P., Chang, K. J. et al. NF-kappaB-dependent induction of microRNA miR-146, an inhibitor targeted to signaling proteins of innate immune responses. *Proc. Natl. Acad. Sci.* U S A. 2006;103:12481-12486.

Takamizawa, J., Konishi, H., Yanagisawa, K. et al. Reduced expression of the *Let-7* microRNAs in human lung cancers in association with shortened postoperative survival. *Cancer Res.* 2004;64:3753-3756.

Thomson, J. M., Parker, J., Perou, C. M. et al. A custom microarray platform for analysis of microRNA gene expression. *Nat. Methods.* 2004;1:47-53.

Thuault, S., Tan, E. J., Peinado, H. et al. HMGA2 and Smads co-regulate SNAIL1 expression during induction of epithelial-to-mesenchymal transition. *J. Biol. Chem.* 2008;283:33437-33446.

Thuault, S., Valcourt, U., Petersen, M. et al. Transforming growth factor-beta employs HMGA2 to elicit epithelial-mesenchymal transition. *J. Cell Biol.* 2006;174:175-183.

Tsang, W. P., Kwok, T. T. *Let-7*a microRNA suppresses therapeutics-induced cancer cell death by targeting caspase-3. *Apoptosis*. 2008;13:1215-1222.

Tsuda, N., Kawano, K., Efferson, C. L. et al. Synthetic microRNA and double-stranded RNA targeting the 3'-untranslated region of HER-2/neu mRNA inhibit HER-2 protein expression in ovarian cancer cells. *Int. J. Oncol.* 2005;27:1299-1306.

Urbich, C., Kuehbacher, A., Dimmeler, S. Role of microRNAs in vascular diseases, inflammation, and angiogenesis. *Cardiovasc. Res.* 2008;79:581-588.

Vasudevan, S., Tong, Y., Steitz, J. A. Switching from repression to activation: microRNAs can up-regulate translation. *Science*. 2007;318:1931-1934.

Vinuesa, C. G., Cook, M. C., Angelucci, C. et al. A RING-type ubiquitin ligase family member required to repress follicular helper T cells and autoimmunity. *Nature*. 2005;435:452-458.

Wang, Q., Li, Y. C., Wang, J. et al. miR-17-92 cluster accelerates adipocyte differentiation by negatively regulating tumor-suppressor Rb2/p130. *Proc. Natl. Acad. Sci.* U S A. 2008;105:2889-2894.

Weidhaas, J. B., Babar, I., Nallur, S. M. et al. MicroRNAs as potential agents to alter resistance to cytotoxic anticancer therapy. *Cancer Res*. 2007;67:11111-11116.

Weiler, J., Hunziker, J., Hall, J. Anti-miRNA oligonucleotides (AMOs): ammunition to target miRNAs implicated in human disease? *Gene Ther*. 2006;13:496-502.

Wulczyn, F. G., SmiRnova, L., Rybak, A. et al. Post-transcriptional regulation of the *Let-7* microRNA during neural cell specification. *FASEB J.* 2007;21:415-426.

Xiao, C., Rajewsky, K. MicroRNA control in the immune system: basic principles. *Cell*. 2009;136:26-36.

Yanaihara, N., Caplen, N., Bowman, E. et al. Unique microRNA molecular profiles in lung cancer diagnosis and prognosis. *Cancer Cell*. 2006;9:189-198.

Yang, H., Kong, W., He, L. et al. MicroRNA expression profiling in human ovarian cancer: miR-214 induces cell survival and cisplatin resistance by targeting PTEN. *Cancer Res*. 2008;68:425-433.

Yang, N., Kaur, S., Volinia, S. et al. MicroRNA microarray identifies *Let-7*i as a novel biomarker and therapeutic target in human epithelial ovarian cancer. *Cancer Res*. 2008;68:10307-10314.

Yang, Z., Bian, C., Zhou, H. et al. MicroRNA hsa-miR-138 inhibits adipogenic differentiation of human adipose tissue-derived mesenchymal stem cells through adenovirus EID-1. *Stem Cells Dev*. 2011;20:259-267.

Yu, D., Tan, A. H., Hu, X. et al. Roquin represses autoimmunity by limiting inducible T-cell co-stimulator messenger RNA. *Nature*. 2007;450:299-303.

Yu, F., Yao, H., Zhu, P. et al. *Let-7* regulates self renewal and tumorigenicity of breast cancer cells. *Cell*. 2007;131:1109-1123.

Zhu, S., Si, M. L., Wu, H. et al. MicroRNA-21 targets the tumor suppressor gene tropomyosin 1 (TPM1). *J. Biol. Chem*. 2007;282:14328-14336.

In: MicroRNA *Let-7*
Editor: Neetu Dahiya

ISBN: 978-1-62081-152-8

Chapter 8

LET-7 AND CANCER

Cesar Seigi Fuziwara, Murilo Vieira Geraldo and Edna Teruko Kimura[1,*]

[1]Department of Cell and Developmental Biology, Institute of Biomedical Sciences, University of São Paulo, São Paulo, Brazil

1. ABSTRACT

In cancer, microRNAs dysregulation affects the post-transcriptional control of oncogenes and tumor suppressors. *Let-7* importance emerged in cancer when its implication in regulating the expression of RAS oncogene family was uncovered. Frequently reduced in several types of cancer, *Let-7* exerts a tumor suppressor role in tumorigenesis, and loss of its expression is associated with poor prognosis. The molecular mechanisms of *Let-7* down-regulation involve genetic alterations such as deletions and loss of heterozygosity, and more recently, post-transcriptional regulation by LIN28, which impairs *Let-7* miRNA maturation. The biological role of *Let-7* is being elucidated gradually by the validation of its targets in relevant pathways such as cyclins, CDKs, E2F2 and MYC in cell cycle; RAS in MAPK pathway; IGF2BP1 in drug resistance; FAS and CASP3 in apoptotic process. As the functional understanding of *Let-7* targets improves, new insights into *Let-7* regulatory network and its potential implication in therapeutics arise. Restoring *Let-7* levels *in vitro* leads to growth restrain and apoptosis of cancer cells, suggesting that in the future therapeutical replacement of *Let-7* could improve or potentiate cancer patient response to common treatment. Thus, unraveling molecular *Let-7* network in cancer is essential to address novel perspectives to translational medicine.

* Corresponding author: Edna Teruko Kimura. Department of Cell and Developmental Biology, Institute of Biomedical Sciences, University of São Paulo, Av. Prof. Lineu Prestes 1524, CEP: 05508-000, São Paulo, SP, Brazil. E-mail: etkimura@usp.br.

2. INTRODUCTION

Initially discovered in *C. elegans* (Pasquinelli, et al., 2000; Reinhart, et al., 2000), *Let-7* microRNA rapidly acquired importance when a clinical study revealed that reduced expression of *Let-7* in lung cancer was related to poor prognosis (Takamizawa et al., 2004). Implied functional consequences of *Let-7* deregulation were revealed by computational screening of *Let-7* family complementary sites in *C. elegans* genes, revealing a high score for putative binding on an ortholog of human RAS, *let-60*, offering validation of the status of RAS family oncogenes as *Let-7* targets (Johnson et al., 2005).

The *Let-7* miRNA family is comprised of 13 genes (*Let-7a* to *Let-7i*, *miR-98*, and *miR-202*) located on different human chromosomes (Roush and Slack, 2008). A bioinformatic search using the TargetScan tool (http://targetscan.org) revealed more than 800 mRNAs predicted to be regulated by *Let-7* family members in humans. These 800 mRNAs represent an array of genes related to different signaling pathways involved in cancer such as LIN28B (Lin28-homolog B), DNA2 (DNA replication 2 homolog), IGFR1 (Insulin-like growth factor 1 receptor), CCNJ (Cyclin J), MAP4K3 (Mitogen-activated protein kinase kinase kinase kinase 3), PIK3IP1 (Phosphoinositide-3-kinase interacting protein 1). However, among numerous predicted mRNA targets, only a few have been validated in functional studies.

High-throughput miRNA screening in broad panels of tumors revealed widespread miRNA deregulation and a characteristic miRNA expression signature during cancer (Lu et al., 2005; Volinia et al., 2006) that could accurately distinguish tumoral from normal tissues. Functional classification of miRNAs has lead to the nomenclature of "oncomiRs" (Esquela-Kerscher and Slack, 2006) for those miRNAs for which altered expression influences cancer through deregulation of oncogenes and tumor suppressor genes (Figure 1). The literature on oncomiRs has grown extensively lately, and the *Let-7* family, together with *miR-21*(Krichevsky and Gabriely, 2009) and the *miR-17-92* cluster (Mendell, 2008), compose the group of miRNAs most commonly altered in cancer. The functional implications of these miRNAs are being uncovered in target validation studies.

3. *LET-7* EXPRESSION IN HUMAN CANCER

In silico screening of miRNA genes in the genome revealed that more than half of 186 miRNA genes analyzed in this study are situated in cancer-associated genomic regions or fragile sites (Calin et al., 2004). *Let-7*family gene loci are located in regions frequently deleted in cancer, such as *Let-7g* on chromosome 3, *Let-7d*, *Let-7a*, and *Let-7f* on chromosome 9, and *Let-7c* on chromosome 21. Accordingly, the presence of *Let-7* in fragile sites could explain, at least partially, the broad reduction of *Let-7* expression observed in different cancers such as lung, ovary, and thyroid (Table 1). On the other hand, up-regulation of *Let-7* has been observed more recently in head, neck, and cervical carcinomas. Overall, the prevalent reduction of *Let-7* oncomiR in cancer suggests a tumor suppressor gene function for this small RNA.

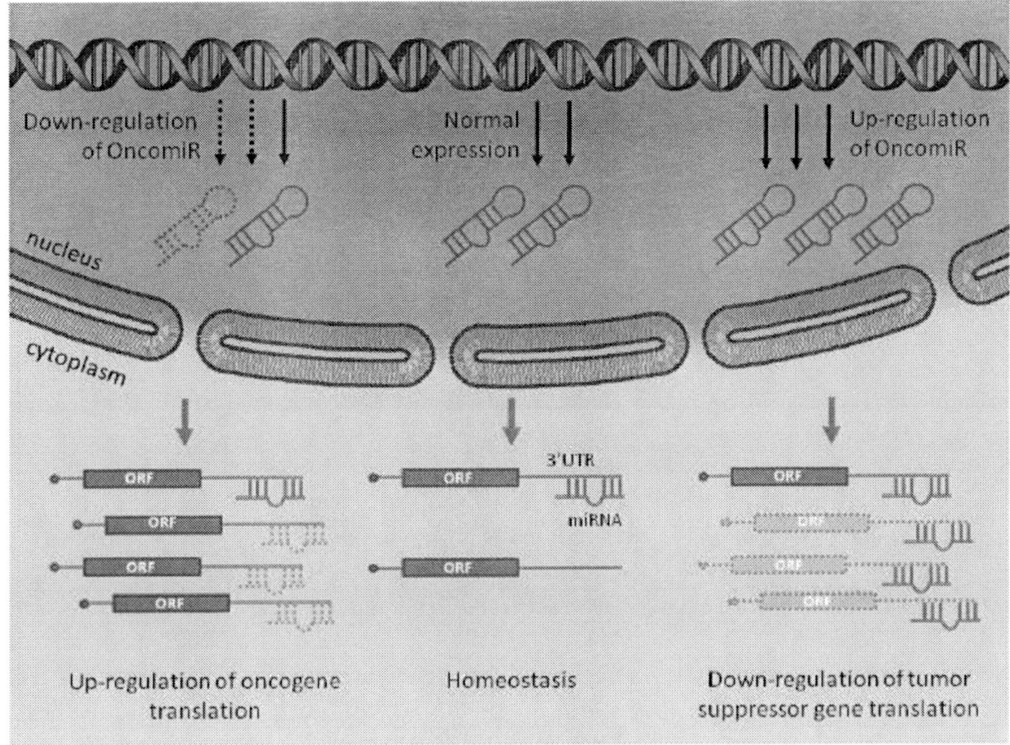

Figure 1. MiRNAs as OncomiRs: Deregulation of miRNAs affects the availability of oncogenic and tumor suppressor proteins by disrupting the repression of translation exerted by miRNAs, leading to enhanced oncogenesis, cell transformation, and proliferation. Dashed lines and arrows represent loss of expression.

A major breakthrough in the miRNA research field in cancer was the seminal study of Johnson et al. (2008) which established the correlation between *Let-7* miRNA and RAS protein levels. The authors demonstrated post-transcriptional regulation of different isoforms of oncogene RAS by binding of *Let-7* in the 3'- untranslated region (3`-UTR) of mRNAs. The possibility of miRNA-mediated control of oncogenes and tumor suppressor genes broadened the functional implications of the role of miRNA in post-transcriptional regulation in general, and miRNA deregulation in cancer etiopathogeny specifically.

1) *Let-7* Down-Regulation in Human Cancer

The first correlation of altered expression of *Let-7* with clinical outcome in human malignancies was reported in lung cancer. In this pioneer study, Takamizawa et al. (2004) reported the down-regulation of *Let-7* levels in non-small cell lung cancer (NSCLC). A marked reduction of more than 80% of *Let-7* expression was observed both in cell lines (60%, 12 out of 20) and NSCLC tumors (44%, 7 out of 16), highlighting *Let-7* as a biological marker in lung tumorigenesis. Moreover, depleted *Let-7a* and *Let-7f* levels were correlated with poorer outcome in NSCLC patients, being associated with advanced disease stages and reduced postoperative survival.

Table 1. *Let-7* is deregulated in different types of cancer

Cancer Type	Sample	*Let-7* Family Expression	Reference
Lung Cancer	Tumor and cell line	Decreased *Let-7* in tumors Decreased *lLet-7* in cancer cel line	Takamizava et al., 2004
	Tumor	Decreased *Let-7a* and *Let-7f* in tumors	Yanaihara et al., 2006
Ovarian Cancer	Tumor	Decreased *Let-7a*, *Let-7c* and *Let-7d* in tumors	Iorio et al., 2007
	Tumor and cell line	Decreased *Let-7b* and *Let-7f* in tumors Decreased *Let-7a*, *Let-7d* and *Let-7f* in cell lines	Dahiya et al., 2008
	Tumor	Decreased *Let-7b*, *Let-7c*, *Let-7d*, *Let-7e*, *Let-7f* and *Let-7i* in tumors	Helland et al., 2011
Prostate Cancer	Tumor and cell line	Decreased *Let-7a* in tumors Decreased *Let-7a* cancer cell line	Dong et al., 2010
	Tumor	Decreased *Let-7c* in tumors	Visone et al., 2007
Thyroid cancer	Tumor	Decreased *Let-7f* in tumors	Pallante et al., 2006
	Tumor	Decreased *Let-7c* in tumors	Visone et al., 2007
Head and Neck cancer	Tumor	Decreased *Let-7a*, *Let-7d* and *Let-7f* in tumors	Childs et al., 2009
	Tumor and metastasis	Decreased *Let-7a* in tumors Decreased *Let-7a* in metastasis	Yu et al., 2011
Gastric Cancer	Tumor, metastasis and cell line	Decreased *Let-7f* in tumors Decreased *Let-7f* in metastasis Decreased *Let-7f* in cancer cell line	Liang et al., 2011
Colon cancer	Tumor	Decreased *Let-7a* and *Let-7b* in tumors	King et al., 2011
Melanoma	Tumor	Decreased *Let-7a*, *Let-7b*, *Let-7d*, *Let-7e* and *Let-7g* in tumors	Schults et al., 2008
	Tumor and cell line	Decreased *Let-7a* in tumors Decreased *Let-7a* in cell lines	Muller et al., 2008
Head and Neck Cancer	Tumor	Increased *Let-7i* in tumors	Chang et al., 2008
Cervical carcinoma	Tumor	Increased *Let-7e*, *Let-7i* and *Let-7g* in tumors	Lee et al., 2008

Studies involving the modulation of *Let-7* miRNA *in vitro* show its importance in cancer cell biology. *Let-7g* repositioning in NSCLC cell lines impairs proliferation and promotes cell death *in vitro*, and a pronounced effect of the blockage of tumorigenesis is observed *in vivo* using NSCLC cells (Kumar et al., 2008). In animal models, *Let-7* strongly represses lung cancer development (~90% reduction in hyperplasia) in *LSL-Kras G12D* transgenic mice (Esquela-Kerscher et al., 2008). Moreover, *Let-7* repositioning renders lung cancer cells sensitive to radiation therapy, decreasing malignant cell survival in a clonogenic assay (Weidhaas et al., 2007). Another important aspect of *Let-7* regulation is exerted by LIN28. LIN28 is an RNA-binding protein that inhibits the maturation of *Let-7* miRNA precursors, therefore disrupting *Let-7* target repression. In small cell lung cancer (SCLC), high levels of LIN28 protein have been associated with depleted *Let-7* expression (Pan et al., 2011). Moreover, small interfering RNA (siRNA) for LIN28 in SCLC cells results in an increase in *Let-7g* mature miRNA and arrest at the G1 phase of the cell cycle.

The systematic study of miRNA in prostate cancer, the most frequent cancer in males, reveals a widespread down-regulation (76 of 85) of commonly expressed miRNAs in normal and tumoral tissues (Ozen et al., 2008), deregulation previously observed in large-scale studies in solid tumors (Lu et al., 2005; Volinia et al., 2006). A marked reduction in the expression of *Let-7* family miRNA (*Let-7b*, *Let-7c*, *Let-7d*, *Let-7f*, *Let-7g*, *Let-7i*), observed by Ozen et al. (2008) highlights the influence of *Let-7* in prostate tumorigenesis. Functional analyses of the role of *Let-7* in the prostate have shown that reintroduction of *Let-7* in cancer cell lines reduces cell proliferation and colony formation *in vitro*, and tumor growth in a xenotransplant model (Dong et al., 2010). Moreover, over-expression of the cell cycle enhancer proteins E2F2 and CCND2, usually present in prostate cancer cell lines, is associated with *Let-7* down-regulation, whichcontrols these proteins level.

Let-7 family expression is reduced in ovarian cancer, indicating that miRNA modulation may enhance ovary tumorigenesis (Iorio et al., 2007). Ovarian cancer displays highly aggressive behavior associated with a marked decrease in *Let-7* in high grade tumors (Dahiya et al., 2008). Moreover, impairment of *Let-7* regulation should contribute to the resistance to chemotherapy that is commonly observed in ovarian cancer, and to the 140,000 deaths yearly worldwide (Jemal et al., 2011). In a recent study, authors report that loss of *Let-7* is associated with resistance to Taxol-based therapy through derepression of IGF2BP1 (insulin-like growth factor 2 mRNA binding protein 1), an RNA binding protein that stabilizes ABCB1 [ATP-binding cassette, sub-family B (MDR/TAP), member 1], a membrane ATP-binding protein cassette that pumps the drug into the extracellular space, consequently inducing refractoriness to chemotherapy (Boyerinas et al., 2011).

An inverse correlation between *Let-7* levels and HMGA2 (an oncoprotein regulated by *Let-7*) protein levels is also observed in digestive system cancers. In esophageal cancer, depleted expression of *Let-7* is observed in cancer tissues, associated with high levels of HMGA2. In contrast, introduction of a *Let-7* mimic in a cancer cell line represses HMGA2 and reduces cell viability (Liu et al., 2011). In gastric cancer, HMGA2 is highly expressed in tumors, and thus, low *Let-7* levels correlate with clinical outcome and survival (Motoyama et al., 2008). Moreover, *Let-7f* reduction has been observed in gastric cancer and metastasis, and functional analysis shows that *Let-7f* introduction in cell line reduces invasive and metastatic potential *in vitro* and *in vivo* by targeting MYH9 (myosin, heavy chain 9, non-muscle) (Liang et al., 2011). In colon cancer, reduced levels of *Let-7b* miRNA are associated with elevation in LIN28B protein level, an inhibitor of *Let-7* miRNA maturation (King et al., 2011). Overall, these studies indicate that *Let-7* influences gastrointestinal tract pathogenesis.

In papillary thyroid carcinoma, the most frequent histotype of thyroid cancer, miRNA studies revealed up-regulation of several miRNAs, such as *miR-146* and *miR-221* and down-regulation of *Let-7f* (He et al. 2005a; Pallante et al. 2006). In normal thyroid cells, the induction of rearrangement of RET/PTC, an upstream activator of RAS in the MAPK pathway, markedly decreased *Let-7f* levels (Ricarte-Filho et al. 2009). Moreover, the functional implication of *Let-7f*-reduced expression in papillary thyroid cancer was explored by inducing *Let-7f* expression in a thyroid cancer cell line, which resulted in reduction in ERK phosphorylation and re-expression of thyroid-specific genes.

Head and neck squamous cell carcinoma (HNSCC) includes tumors of the oral cavity, oropharynx, and larynx, with an estimated 5-year survival period in 50% of patients. High-throughput screening for miRNA deregulation in HNSCC has provided important advances in molecular research for molecular markers, and miRNA deregulation seems to be an important step in tumorigenesis. Microarray analysis has revealed *Let-7* family down-regulation in HNSCC (Childs et al., 2009). In this study, low *Let-7d* combined with low *miR-205* miRNA expression was associated with overall increased risk, loco-regional recurrence, and decreased survival as cancer progressed. Moreover, another study showed a stronger reduction in *Let-7a* miRNA in metastatic and recurrent HNSCC, and found an association between *Let-7* repression and reactivation of characteristic stem cell-like markers in cancer cells prone to drug resistance (Yu et al., 2011).

2) *Let-7* Up-Regulation in Human Cancer

Although not so frequent, *Let-7* up-regulation has been reported in different types of cancer. In cervical cancer, a gynecological malignancy, miRNA profile analysis in a set of early invasive squamous cell carcinoma tumors shows large scale up-regulation of several miRNAs including *Let-7e*, *Let-7i*, and *Let-7g* (Lee et al., 2008), indicating an important influence of *Let-7* in cervical cancer. The analysis of rare but aggressive anaplastic thyroid cancer revealed increased expression of *Let-7a* and *Let-7f* isoforms (Visone et al., 2007), suggesting that deregulation of *Let-7* isoforms participates in thyroid cancer oncogenesis. In head and neck cancer, increased *Let-7i* expression was detected in a microarray-based study of several miRNAs from HNSCC primary tissue (Chang et al., 2008); however, the functional implication of this remains to be explored. Thus, *Let-7* upregulation also occurs in cancer, indicating that *Let-7* may play a different role depending on cancer histotype and origin.

4. *LET-7* TARGETS IN HUMAN CANCER

The functional role of *Let-7* in cancer has been investigated in diverse studies using cancer cells and molecular approaches to inhibit or over-express *Let-7*. However, from hundreds of predicted mRNA targets of the *Let-7* family, only about a dozen have been functionally validated, indicating that the next step after reporting results from miRNA deregulation studies will be exploring their biological and functional implications in cancer.

So far, the validated targets of *Let-7* include proteins related to apoptosis, cell proliferation, cell cycle regulation, and resistance to drug treatment (Figure 2).

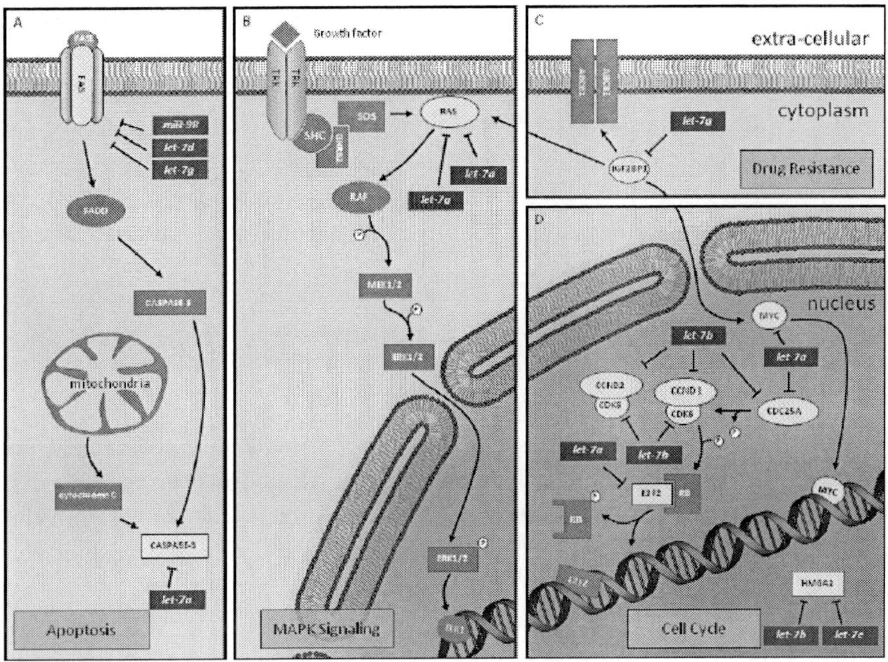

Figure 2. *Let-7*-mediated regulation network. Various signaling pathways are repressed by *Let-7* family miRNAs and deregulation leads to complex and intricate patterns of alteration in levels of cancer suppressor and promoter proteins. A, Apoptosis pathway; B, MAPK signaling; C, Drug resistance; D, Cell cycle regulation. White boxes and balloons contain validated *Let-7* targets, and black boxes contain *Let-7* isoforms.

1) HMGA2

High Mobility Group AT-hook2 (HMGA2) is a member of the non-histone chromatin-associated proteins, which are architectural factors essential for formation of the enhanceosome. HMGA2 acts as an oncofetal protein which is expressed in embryonic tissues, participating in the maintenance of pluripotency (Boyerinas et al., 2008), but it is not detected in normal adult tissues. High HMGA2 protein levels have been detected in various human neoplasias; for example, in 90% of lung cancers associated with poor prognosis. Chromosomal rearrangements of the HMGA2 gene have been reported in several types of cancer, resulting in over-expression of protein.

The chromosomal break separates the coding region of the gene from its 3`-UTR and fuses it with another gene from the translocated region. This characteristic feature of rearrangement indicates that the loss of 3`-UTR could be an important step in tumorigenesis (Fusco and Fedele, 2007). Indeed, the 3`-UTR is important for post-transcriptional regulation exerted by microRNAs, especially for the *Let-7* miRNA family, which has five predicted sites of binding in the HMGA2 3`-UTR. An inverse correlation between *Let-7* and HMGA2 expression has been observed in lung cancer patients, and the introduction of exogenous *Let-7* rescues growth inhibition of lung cancer cells, indicating that *Let-7* may influence lung tumorigenesis (Lee and Dutta, 2007). Moreover, by disrupting the pairing of *Let-7* with the HMGA2 3' UTR, an increase in HMGA2 protein and tumorigenicity was observed *in vivo,* pointing to the loss of miRNA-directed regulation as a predisposing factor to cancer development (Mayr et al., 2007).

Besides HMGA2, another oncofetal protein regulated by *Let-7* is IGF2BP1 (insulin-like growth factor 2 mRNA binding protein 1), expressed in the oncofetal period and which regulates protein degradation, prolonging the half-life of IGF2 mRNA, MYC, KRAS, and other genes (Mongroo et al., 2011; Nielsen et al., 1999; Noubissi et al., 2006), as well as stimulating proliferation. This intricate regulatory network indicates that *Let-7* is essential for the maintenance of differentiated status.

2) RAS Family

The RAS oncogene was the first validated human target of the *Let-7* family. Johnson et al. (2005) demonstrated that the 3`-UTR of different RAS isoforms contained multiple predicted binding sites for *Let-7* miRNAs. RAS family proteins signal through the MAPK pathway to enhance proliferation and survival (Figure 2B). However, RAS can also cross-talk to different pathways such as AKT and PI3Kinase. Activating mutations in RAS genes are detected in about 15% of cancers, comprising different histological origins such as lung, colorectal, pancreatic, and thyroid cancer among others (Lau and Haigis, 2009). *Let-7* down-regulation enhances cell transformation by RAS oncogene activation and predicts poor prognosis as observed in lung cancer (Takamizawa et al., 2004). Moreover, *Let-7g* potently represses KRAS and NRAS protein levels in a lung cancer cell line, reducing cell proliferation and inducing death, and also markedly inhibiting tumorigenicity *in vivo* (Kumar et al., 2008).

Further analysis of the RAS gene shed light in the importance of the integrity of the 3'-UTR region in gene regulation. Johnson et al. reported the existence of several predicted imperfect pairing sites for *Let-7* in the 3'-UTRs of RAS isoforms such as HRAS, KRAS, and NRAS (3, 8, and 9 predicted binding sites, respectively) indicating the importance of *Let-7*-mediated control over RAS. Recently, studies have pointed to the importance of single nucleotide polymorphisms (SNP) in the 3'-UTRs of RAS isoforms that impair *Let-7* binding. In lung cancer, the occurrence of the LCS6 SNP in the KRAS 3'-UTR predicts increased risk of developing lung cancer in moderate smokers (Chin et al., 2008), probably through the disruption of *Let-7*-mediated inhibition of KRAS. This SNP is detected in over 18% of lung cancer cases, but its prevalence in the overall population is about 5%. Moreover, the presence of LCS6 is associated with reduced survival of patients with oral cancer (Christensen et al., 2009), indicating that not only *Let-7* miRNA levels, but also the integrity of binding sites (absence of SNPs) which interfere with miRNA pairing and protein regulation, are important as prognostic factors.

3) MYC

MYC, another oncogene regulated by *Let-7*, is a nuclear protein that interacts with MAD to form a heterocomplex that acts as a transcription factor and regulates cell proliferation, apoptosis, metabolism, and differentiation (Wasylishen and Penn, 2010) (Figure 2D). Oncogenic activation of MYC is observed in different types of tumors, such as Burkitt lymphoma, osteosarcoma, glioblastoma, and cervical and breast carcinoma (Facchini and Penn, 1998).

The *Let-7* post-transcriptional regulation of MYC elucidates part of the intricate functional network of *Let-7* (Sampson et al., 2007). In this study, treatment of a Burkitt Lymphoma cell line with a MYC inhibitor significantly increases expression levels of *Let-7a*, *Let-7b*, and *miR-98*. Reposition of *Let-7a* in a MYC-expressing fibroblast cell line reduces MYC mRNA and protein. Moreover, knock-down of MYC by siRNA results in a more than twofold increase in *Let-7a* expression in comparison with MYC-expressing fibroblasts. Interestingly, the same authors also observed down-regulation of *miR-17-5p*, one of the components of the *miR-17-92* cluster. The MYC oncogene activates miRNA expression, specially *miR-17-92*, in a positive feedback loop that results in enhanced tumorigenesis in a mouse model of lymphoma (He et al., 2005b), as well as stimulation of angiogenesis (Dews et al., 2006). Therefore, *Let-7* down-regulation may be an important step in MYC-induced cell proliferation and activation of oncogenic *miR-17-92*.

4) Cell Cycle Related Genes: CCND1, CCND2, CDK6, CDC25A, and E2F2

A lack of balance in cell cycle control and consequent unrestrained growth are the main characteristics of cancer biology. Cell cycle progression in the G1 phase is mediated by activation of the Cyclin/CDK complex by CDC25A-dependent phosphorylation of CDK, which in turn phosphorylates downstream targets such as the RB/E2F complex. Hyperphosphorylation of RB disassembles the complex and releases the transcription factor E2F, which is able to activate the transcription of its targets for transition to the S phase of cell cycle.

Let-7 isoforms regulate the translational levels of cyclin D1 (CCND1) (Schultz et al., 2008), cyclin D2 (CCND2), E2F transcription factor 2 (E2F2) (Dong et al., 2010), cyclin-dependent kinase-6 (CDK6), and cell division cycle homolog A (CDC25A) (Johnson et al., 2007), all of which are enhancer proteins of cell cycle progression (Figure 2D). Oncogene activation increases cell cycle-promoting protein levels as it inhibits *Let-7* and disrupts *Let-7*-mediated regulation, ultimately leading to unrestrained growth. Therefore, *Let-7* negatively regulates the proliferation dynamic by modulating the availability of proteins essential to cell cycle entry and progress.

5) Apoptosis-Related Genes: Caspase 3 and FAS

Let-7 targets apoptotic proteins such as Caspase 3 (CASP3) and TNF receptor superfamily, member 6 (FAS) (Figure 2A). CASP3 is the downstream effector protein in the apoptotic process. In hepatocellular and squamous cell carcinoma cells, over-expression of

Let-7a is observed in drug-resistant cell lines. *Let-7a* represses CASP3 protein, leading to resistance to doxorubicin, paclitaxel, and interferon treatment and enhancing cell survival (Tsang and Kwok, 2008), pointing to the involvement of *Let-7* deregulation in refractoriness to chemotherapy.

The cell death-inducing signaling pathway is activated by the binding of FASL to FAS, ultimately triggering caspase-3. Deregulation of FAS signaling is observed in different pathologies. In colon cancer, loss of expression or down-regulation of FAS is usually observed during tumor progression (Moller et al., 1994). A computational search for FAS-associated proteins leads to the discovery of FAS mRNA as a potential target for *Let-7/miR-98* miRNAs. Functional validation in cell line models revealed that *Let-7d*, *Let-7g*, and *miR-98* repressed FAS translation by binding to the 3`-UTR of the mRNA (Wang et al., 2011). Moreover, stimulation of CD4+ T lymphocytes with anti-CD23 and anti-CD28 antibodies, in a mechanism known as activation-induced cell death, leads to a reduction in the time course of *Let-7-g* expression which correlates with increasing FAS protein levels.

5. *LET-7* IN CANCER SURVIVAL AND DRUG RESISTANCE

Deregulation of *Let-7* expression has been associated with poor prognosis and drug therapy resistance in cancer. *Let-7* family members act predominantly as tumor suppressor genes, controlling cell cycle proteins and cell differentiation. Once *Let-7* is deregulated, malignant cells may acquire stem-like properties, becoming resistant to chemotherapy. This acquired resistance can be associated with over-expression of membrane-bound proteins such as ATP-binding cassette (ABC) proteins, which pump drugs into the extracellular compartment.

In head and neck cancer, reduced *Let-7* levels increase the expression of stem cell markers such as NANOG and OCT4. Moreover, *Let-7a* over-expression renders cancer cells more sensitive to cisplatin treatment which decreases mRNA levels of ABCB1, ABCG2, and ABCG5, genes involved in chemo-resistance (Yu et al. 2011). In ovarian cancer, chemoresistance is also associated with ABCB1, and the introduction of *Let-7g* sensitizes cells to chemotherapy through reduction of expression of IGF2BP1, an RNA binding protein that regulates RNA stability (Boyerinas et al., 2011) (Figure 2C). On the other hand, mimicking drug resistance by treatment of cholangiocarcinoma cells with IL6 (Interleukin-6) significantly increases *Let-7* family miRNA expression, enhances cell survival, and reduces apoptosis via repression of the transcription factor NF2, which leads to up-regulation of STAT3 phosphorylation (Meng et al., 2007).

6. *LET-7* MATURATION CONTROL BY LIN28

Besides the intrinsic regulatory mechanisms underlying miRNA transcription and processing reported in other section, an emerging role has been uncovered for LIN28. LIN28 homolog or LIN28, a predicted target for *Let-7*, is an RNA binding protein that possesses two RNA binding domains: a "cold shock" domain (CSD) and a pair of retroviral-type CCHC zinc-finger domains (Moss and Tang 2003). Emerging data point to an important role for

LIN28 in controlling *Let-7* mature miRNA expression, in a regulatory loop that involves repression of the processing of precursors to mature *Let-7* miRNA. Reduced expression of *Let-7* family miRNAs in cancer can be associated with over-expression of LIN28, which has been reported in different types of human malignancies (Viswanathan et al., 2009). In this study, the authors showed that LIN28 over-expression promotes malignant transformation of fibroblasts by depleting *Let-7* miRNA levels and derepressing *Let-7* targets. Accordingly, LIN28 knockdown resulted in growth impairment and induction of differentiation in cancer cells. Moreover, *Let-7* expression is reduced or almost absent in a variety of stem cell populations and progenitor cells (Richards et al., 2004). Interestingly, LIN28 induces pluripotency in human fibroblasts, acting together with NANOG, SOX2, and KLF4 to recover pluripotency in somatic cells (Yu et al., 2007).

Thus, interference in post-transcriptional regulation of mature *Let-7* miRNA levels by LIN28 may be an additional mechanism for transformation, complementing the effects triggered by oncogene activation.

Conclusion

Understanding the mechanism underlying the escape of cancer cells from miRNA post-transcriptional regulation, points to a further level of complexity in cancer research. Unbalanced *Let-7* expression seems to be a key event during oncogenic transformation, as majority of cancers display a reduction in the level of *Let-7* family expression. Disruption of *Let-7*-mediated regulation may be the first step in tumorigenesis, as *Let-7* exerts a potent, complex, and intricate regulatory effect on different targets, mainly controlling cell cycle entry, proliferation, and differentiation pathways. As studies progress, further questions are raised about how miRNAs can interfere with the fine tuning of cell homeostasis and, through this, regulation of various genes. Overall, modulation of *Let-7* miRNAs could, in the future, lead to new therapeutic approaches for the treatment of more aggressive and chemotherapy-resistant cancers, in an attempt to improve patient survival by restoring miRNA-based regulation of oncoproteins.

References

Boyerinas, B., Park, S. M., Murmann, A. E., Gwin, K., Montag, A. G., Zillardt, M. R., Hua, Y. J., Lengyel, E., and Peter, M. E. *Let-7* modulates acquired resistance of ovarian cancer to Taxanes via IMP-1-mediated stabilization of MDR1. *Int. J. Cancer*. 2011.

Boyerinas, B., Park, S. M., Shomron, N., Hedegaard, M. M., Vinther, J., Andersen, J. S., Feig, C., Xu, J., Burge, C. B., and Peter, M. E. Identification of *Let-7*-regulated oncofetal genes. *Cancer Res.* 2008;68:2587-2591.

Calin, G. A., Sevignani, C., Dumitru, C. D., Hyslop, T., Noch, E., Yendamuri, S., Shimizu, M., Rattan, S., Bullrich, F., Negrini, M., and Croce, C. M. Human microRNA genes are frequently located at fragile sites and genomic regions involved in cancers. *Proc. Natl. Acad. Sci.* U S A. 2004;101:2999-3004.

Chang, S. S., Jiang, W. W., Smith, I., Poeta, L. M., Begum, S., Glazer, C., Shan, S., Westra, W., Sidransky, D., and Califano, J. A. MicroRNA alterations in head and neck squamous cell carcinoma. *Int. J. Cancer*. 2008;123:2791-2797.

Childs, G., Fazzari, M., Kung, G., Kawachi, N., Brandwein-Gensler, M., McLemore, M., Chen, Q., Burk, R. D., Smith, R.V., Prystowsky, M. B., Belbin, T. J., and Schlecht, N. F. Low-level expression of microRNAs *Let-7*d and miR-205 are prognostic markers of head and neck squamous cell carcinoma. *Am. J. Pathol*. 2009;174:736-745.

Chin, L. J., Ratner, E., Leng, S., Zhai, R., Nallur, S., Babar, I., Muller, R. U., Straka, E., Su, L., Burki, E. A., Crowell, R. E., Patel, R., Kulkarni, T., Homer, R., Zelterman, D., Kidd, K. K., Zhu, Y., Christiani, D. C., Belinsky, S. A., Slack, F. J., and Weidhaas, J. B. A SNP in a *Let-7* microRNA complementary site in the KRAS 3' untranslated region increases non-small cell lung cancer risk. *Cancer Res*. 2008;68:8535-8540.

Christensen, B. C., Moyer, B. J., Avissar, M., Ouellet, L. G., Plaza, S. L., McClean, M. D., Marsit, C. J., and Kelsey, K. T. A *Let-7* microRNA-binding site polymorphism in the KRAS 3' UTR is associated with reduced survival in oral cancers. *Carcinogenesis*. 2009;30:1003-1007.

Dahiya, N., Sherman-Baust, C. A., Wang, T. L., Davidson, B., Shih Ie, M., Zhang, Y., Wood, W., 3rd, Becker, K. G., and Morin, P. J. MicroRNA expression and identification of putative miRNA targets in ovarian cancer. *PLoS One*. 2008;3:e2436.

Dews, M., Homayouni, A., Yu, D., Murphy, D., Sevignani, C., Wentzel, E., Furth, E. E., Lee, W. M., Enders, G. H., Mendell, J. T., and Thomas-Tikhonenko, A. Augmentation of tumor angiogenesis by a Myc-activated microRNA cluster. *Nat. Genet*. 2006;38:1060-1065.

Dong, Q., Meng, P., Wang, T., Qin, W., Qin, W., Wang, F., Yuan, J., Chen, Z., Yang, A., and Wang, H. MicroRNA *Let-7*a inhibits proliferation of human prostate cancer cells in vitro and in vivo by targeting E2F2 and CCND2. *PLoS One*. 2010;5:e10147.

Esquela-Kerscher, A. and Slack, F. J. OncomiRs - microRNAs with a role in cancer. *Nat. Rev. Cancer*. 2006;6:259-269.

Esquela-Kerscher, A., Trang, P., Wiggins, J. F., Patrawala, L., Cheng, A., Ford, L., Weidhaas, J. B., Brown, D., Bader, A. G., and Slack, F. J. The *Let-7* microRNA reduces tumor growth in mouse models of lung cancer. *Cell Cycle*. 2008;7:759-764.

Facchini, L. M. and Penn, L. Z. The molecular role of Myc in growth and transformation: recent discoveries lead to new insights. *Faseb. J.* 1998;12:633-651.

Fusco, A. and Fedele, M. Roles of HMGA proteins in cancer. *Nat. Rev. Cancer*. 2007;7:899-910.

He, H., Jazdzewski, K., Li, W., Liyanarachchi, S., Nagy, R., Volinia, S., Calin, G. A., Liu, C. G., Franssila, K., Suster, S., Kloos, R. T., Croce, C. M., and De la Chapelle, A. The role of microRNA genes in papillary thyroid carcinoma. *Proc. Natl. Acad. Sci.* U S A. 2005a;102:19075-19080.

He, L., Thomson, J. M., Hemann, M. T., Hernando-Monge, E., Mu, D., Goodson, S., Powers, S., Cordon-Cardo, C., Lowe, S. W., Hannon, G. J., and Hammond, S. M. A microRNA polycistron as a potential human oncogene. *Nature*. 2005b;435:828-833.

Iorio, M. V., Visone, R., Di Leva, G., Donati, V., Petrocca, F., Casalini, P., Taccioli, C., Volinia, S., Liu, C. G., Alder, H., Calin, G. A., Menard, S., and Croce, C. M. MicroRNA signatures in human ovarian cancer. *Cancer Res*. 2007;67:8699-8707.

Jemal, A., Bray, F., Center, M. M., Ferlay, J., Ward, E., and Forman, D. Global cancer statistics. *CA Cancer J. Clin.* 2011;61:69-90.

Johnson, C. D., Esquela-Kerscher, A., Stefani, G., Byrom, M., Kelnar, K., Ovcharenko, D., Wilson, M., Wang, X., Shelton, J., Shingara, J., Chin, L., Brown, D., and Slack, F. J. The *Let-7* microRNA represses cell proliferation pathways in human cells. *Cancer Res.* 2007;67:7713-7722.

Johnson, S. M., Grosshans, H., Shingara, J., Byrom, M., Jarvis, R., Cheng, A., Labourier, E., Reinert, K. L., Brown, D., and Slack, F. J. RAS is regulated by the *Let-7* microRNA family. *Cell.* 2005;120:635-647.

King, C. E., Wang, L., Winograd, R., Madison, B. B., Mongroo, P. S., Johnstone, C. N., and Rustgi, A. K. LIN28B fosters colon cancer migration, invasion and transformation through *Let-7*-dependent and -independent mechanisms. *Oncogene.* 2011.

Krichevsky, A. M., and Gabriely, G. miR-21: a small multi-faceted RNA. *J. Cell Mol. Med.* 2009;13:39-53.

Kumar, M. S., Erkeland, S. J., Pester, R. E., Chen, C. Y., Ebert, M. S., Sharp, P. A., and Jacks, T. Suppression of non-small cell lung tumor development by the *Let-7* microRNA family. *Proc. Natl. Acad. Sci.* U S A. 2008;105:3903-3908.

Lau, K. S. and Haigis, K. M. Non-redundancy within the RAS oncogene family: insights into mutational disparities in cancer. *Mol. Cells.* 2009;28:315-320.

Lee, J. W., Choi, C. H., Choi, J. J., Park, Y. A., Kim, S. J., Hwang, S. Y., Kim, W. Y., Kim, T. J., Lee, J. H., Kim, B. G., and Bae, D. S. Altered MicroRNA expression in cervical carcinomas. *Clin. Cancer Res.* 2008;14:2535-2542.

Lee, Y. S. and Dutta, A. The tumor suppressor microRNA *Let-7* represses the HMGA2 oncogene. *Genes Dev.* 2007;21:1025-1030.

Liang, S., He, L., Zhao, X., Miao, Y., Gu, Y., Guo, C., Xue, Z., Dou, W., Hu, F., Wu, K., Nie, Y., and Fan, D. MicroRNA *Let-7*f inhibits tumor invasion and metastasis by targeting MYH9 in human gastric cancer. *PLoS One.*2011;6:e18409.

Liu, Q., Lv, G. D., Qin, X., Gen, Y. H., Zheng, S. T., Liu, T., and Lu, X. M. Role of microRNA *Let-7* and effect to HMGA2 in esophageal squamous cell carcinoma. *Mol. Biol. Rep.* 2011.

Lu, J., Getz, G., Miska, E. A., Alvarez-Saavedra, E., Lamb, J., Peck, D., Sweet-Cordero, A., Ebert, B. L., Mak, R. H., Ferrando, A. A., Downing, J. R., Jacks, T., Horvitz, H. R., and Golub, T. R. MicroRNA expression profiles classify human cancers. *Nature.* 2005;435:834-838.

Mayr, C., Hemann, M. T. and Bartel, D. P. Disrupting the pairing between *Let-7* and Hmga2 enhances oncogenic transformation. *Science.* 2007;315:1576-1579.

Mendell, J. T. miRiad roles for the miR-17-92 cluster in development and disease. *Cell.* 2008;133:217-222.

Meng, F., Henson, R., Wehbe-Janek, H., Smith, H., Ueno, Y., and Patel, T. The MicroRNA *Let-7*a modulates interleukin-6-dependent STAT-3 survival signaling in malignant human cholangiocytes. *J. Biol. Chem.* 2007;282:8256-8264.

Moller, P., Koretz, K., Leithauser, F., Bruderlein, S., Henne, C., Quentmeier, A., and Krammer, P. H. Expression of APO-1 (CD95), a member of the NGF/TNF receptor superfamily, in normal and neoplastic colon epithelium. *Int. J. Cancer.* 1994;57:371-377.

Mongroo, P. S., Noubissi, F. K., Cuatrecasas, M., Kalabis, J., King, C. E., Johnstone, C. N., Bowser, M. J., Castells, A., Spiegelman, V. S., and Rustgi, A. K. IMP-1 displays crosstalk with K-Ras and modulates colon cancer cell survival through the novel proapoptotic protein CYFIP2. *Cancer Res*. 2011;71:2172-2182.

Moss, E. G. and Tang, L. Conservation of the heterochronic regulator Lin-28, its developmental expression and microRNA complementary sites. *Dev. Biol*. 2003;258:432-442.

Motoyama, K., Inoue, H., Nakamura, Y., Uetake, H., Sugihara, K., and Mori, M. Clinical significance of high mobility group A2 in human gastric cancer and its relationship to *Let-7* microRNA family. *Clin. Cancer Res*. 2008;14:2334-2340.

Nielsen, J., Christiansen, J., Lykke-Andersen, J., Johnsen, A. H., Wewer, U. M., and Nielsen, F. C. A family of insulin-like growth factor II mRNA-binding proteins represses translation in late development. *Mol. Cell Biol*. 1999;19:1262-1270.

Noubissi, F. K., Elcheva, I., Bhatia, N., Shakoori, A., Ougolkov, A., Liu, J., Minamoto, T., Ross, J., Fuchs, S. Y., and Spiegelman, V. S. CRD-BP mediates stabilization of betaTrCP1 and c-myc mRNA in response to beta-catenin signalling. *Nature*. 2006;441:898-901.

Ozen, M., Creighton, C. J., Ozdemir, M., and Ittmann, M. Widespread deregulation of microRNA expression in human prostate cancer. *Oncogene*. 2008;27:1788-1793.

Pallante, P., Visone, R., Ferracin, M., Ferraro, A., Berlingieri, M. T., Troncone, G., Chiappetta, G., Liu, C. G., Santoro, M., Negrini, M., Croce, C. M., and Fusco, A. MicroRNA deregulation in human thyroid papillary carcinomas. *Endocr. Relat. Cancer*. 2006;13:497-508.

Pan, L., Gong, Z., Zhong, Z., Dong, Z., Liu, Q., Le, Y., and Guo, J. Lin-28 reactivation is required for *Let-7* repression and proliferation in human small cell lung cancer cells. *Mol. Cell Biochem*. 2011;355:257-263.

Pasquinelli, A. E., Reinhart, B. J., Slack, F., Martindale, M. Q., Kuroda, M. I., Maller, B., Hayward, D. C., Ball, E. E., Degnan, B., Muller, P., Spring, J., Srinivasan, A., Fishman, M., Finnerty, J., Corbo, J., Levine, M., Leahy, P., Davidson, E., and Ruvkun, G. Conservation of the sequence and temporal expression of *Let-7* heterochronic regulatory RNA. *Nature*. 2000;408:86-89.

Reinhart, B. J., Slack, F. J., Basson, M., Pasquinelli, A. E., Bettinger, J. C., Rougvie, A. E., Horvitz, H. R., and Ruvkun, G. The 21-nucleotide *Let-7* RNA regulates developmental timing in Caenorhabditis elegans. *Nature*. 2000;403:901-906.

Ricarte-Filho, J. C., Fuziwara, C. S., Yamashita, A. S., Rezende, E., Da-Silva, M. J., and Kimura, E. T. Effects of *Let-7* microRNA on Cell Growth and Differentiation of Papillary Thyroid Cancer. *Transl. Oncol*. 2009;2:236-241.

Richards, M., Tan, S. P., Tan, J. H., Chan, W. K., and Bongso, A. The transcriptome profile of human embryonic stem cells as defined by SAGE. *Stem. Cells*. 2004;22:51-64.

Roush, S. and Slack, F. J. The *Let-7* family of microRNAs. *Trends Cell Biol*. 2008;18:505-516.

Sampson, V. B., Rong, N. H., Han, J., Yang, Q., Aris, V., Soteropoulos, P., Petrelli, N. J., Dunn, S. P., and Krueger, L. J. MicroRNA *Let-7*a down-regulates MYC and reverts MYC-induced growth in Burkitt lymphoma cells. *Cancer Res*. 2007;67:9762-9770.

Schultz, J., Lorenz, P., Gross, G., Ibrahim, S., and Kunz, M. MicroRNA *Let-7*b targets important cell cycle molecules in malignant melanoma cells and interferes with anchorage-independent growth. *Cell Res*. 2008;18:549-557.

Takamizawa, J., Konishi, H., Yanagisawa, K., Tomida, S., Osada, H., Endoh, H., Harano, T., Yatabe, Y., Nagino, M., Nimura, Y., Mitsudomi, T., and Takahashi, T. Reduced expression of the *Let-7* microRNAs in human lung cancers in association with shortened postoperative survival. *Cancer Res.* 2004;64:3753-3756.

Tsang, W. P. and Kwok, T. T. *Let-7*a microRNA suppresses therapeutics-induced cancer cell death by targeting caspase-3. *Apoptosis.* 2008;13:1215-1222.

Visone, R., Pallante, P., Vecchione, A., Cirombella, R., Ferracin, M., Ferraro, A., Volinia, S., Coluzzi, S., Leone, V., Borbone, E., Liu, C. G., Petrocca, F., Troncone, G., Calin, G. A., Scarpa, A., Colato, C., Tallini, G., Santoro, M., Croce, C. M., and Fusco, A. Specific microRNAs are downregulated in human thyroid anaplastic carcinomas. *Oncogene.* 2007;26:7590-7595.

Viswanathan, S. R., Powers, J. T., Einhorn, W., Hoshida, Y., Ng, T. L., Toffanin, S., O'Sullivan, M., Lu, J., Phillips, L. A., Lockhart, V. L., Shah, S. P., Tanwar, P. S., Mermel, C. H., Beroukhim, R., Azam, M., Teixeira, J., Meyerson, M., Hughes, T. P., Llovet, J. M., Radich, J., Mullighan, C. G., Golub, T. R., Sorensen, P. H., and Daley, G. Q. Lin28 promotes transformation and is associated with advanced human malignancies. *Nat. Genet.* 2009;41:843-848.

Volinia, S., Calin, G. A., Liu, C. G., Ambs, S., Cimmino, A., Petrocca, F., Visone, R., Iorio, M., Roldo, C., Ferracin, M., Prueitt, R. L., Yanaihara, N., Lanza, G., Scarpa, A., Vecchione, A., Negrini, M., Harris, C. C., and Croce, C. M. A microRNA expression signature of human solid tumors defines cancer gene targets. *Proc. Natl. Acad. Sci.* U S A. 2006;103:2257-2261.

Wang, S., Tang, Y., Cui, H., Zhao, X., Luo, X., Pan, W., Huang, X., and Shen, N. *Let-7*/miR-98 regulate Fas and Fas-mediated apoptosis. *Genes Immun.* 2011;12:149-154.

Wasylishen, A. R. and Penn, L. Z. Myc: the beauty and the beast. *Genes Cancer.* 2010;1:532-541.

Weidhaas, J. B., Babar, I., Nallur, S. M., Trang, P., Roush, S., Boehm, M., Gillespie, E., and Slack, F. J. MicroRNAs as potential agents to alter resistance to cytotoxic anticancer therapy. *Cancer Res.* 2007;67:11111-11116.

Yu, C. C., Chen, Y. W., Chiou, G. Y., Tsai, L. L., Huang, P. I., Chang, C. Y., Tseng, L. M., Chiou, S. H., Yen, S. H., Chou, M. Y., Chu, P. Y., and Lo, W. L. MicroRNA *Let-7*a represses chemoresistance and tumourigenicity in head and neck cancer via stem-like properties ablation. *Oral Oncol.* 2011;47:202-210.

Yu, J., Vodyanik, M. A., Smuga-Otto, K., Antosiewicz-Bourget, J., Frane, J. L., Tian, S., Nie, J., Jonsdottir, G. A., Ruotti, V., Stewart, R., Slukvin, I. I., and Thomson, J. A. Induced pluripotent stem cell lines derived from human somatic cells. *Science.* 2007;318:1917-1920.

In: MicroRNA *Let-7*
Editor: Neetu Dahiya

ISBN: 978-1-62081-152-8

Chapter 9

ROLES OF *LET-7* FAMILY MIRNAS IN DEVELOPMENT AND DIFFERENTIATION

Rikako Sanuki and Takahisa Furukawa[1*]

[1]Department of Developmental Biology, Osaka Bioscience Institute, 6-2-4, Furuedai, Suita, Osaka, Japan

1. ABSTRACT

MiRNAs (miRNAs) are a class of 18 to 25-nucleotide, short, non-coding RNAs that regulate gene expression post-transcriptionally. The lethal-7 (*Let-7*) gene is one of the founding members of the first two miRNA families identified in *C. elegans*, and the first miRNA known to be found in humans. *Let-7* family miRNAs are expressed from early development through maturity in both invertebrates and vertebrates. The majority of miRNAs are not essential for viability or development in *C. elegans*. However, *Let-7* family miRNAs are critical regulators of development that control both early and late developmental timing decisions, and were elucidated by extensive genetic and molecular analysis using *C. elegans* mutants as well as *Drosophila* mutants. In contrast to invertebrates, our understanding of the functional roles of *Let-7* family miRNAs in vertebrate development is still limited because there are many redundant *Let-7* family miRNAs, and thus experiments to clarify *in vivo* functions of *Let-7* family miRNAs using *Let-7* family miRNA mutants in vertebrates are practically difficult. However, indirect but significant *in vivo* evidence has suggested that *Let-7* is essential for gene regulation in mouse development. In addition, ectopic expression of *Let-7* in zebrafish embryos leads to early down-regulation of endogenous *Let-7* targets, resulting in severe embryonic developmental abnormalities, and suggesting that *Let-7* plays a significant role in zebrafish development. Although studies on vertebrate *Let-7* family miRNAs are still in the early stages, these studies have provided us with evidence of important biological roles for *Let-7*. Based on past and recent findings, we review the expression patterns, biological roles, and functional mechanisms of *Let-7* family miRNAs in normal development of both invertebrates and vertebrates. In addition, we describe *in vitro* analysis on the molecular roles of *Let-7* family miRNAs underlying cellular

* Corresponding author: Takahisa Furukava. Department of Developmental Biology, Osaka Bioscience Institute, 6-2-4, Furuedai, Suita, Osaka, 565-0874, Japan, Email: furukawa@obi.or.jp.

differentiation. Future studies on *Let-7* family miRNAs will provide us with not only important clues to understanding cell differentiation in development, but also useful tools for diagnosis and/or therapies in the medical field.

2. Introduction

Development requires an intricate machinery of cell proliferation, differentiation and morphogenesis. Understanding how developmental events are coordinated in developmental schedule needs identification of the genes that regulate developmental timing. The heterochrony, a term that describes changes in the relative timing of developmental event between a mutant and wild-type organism or between distinct species (Pasquinelli and Ruvkun, 2002). Mutations in heterochronic genes cause misregulations of developmental events. In 2000, *Let-7* was identified as a heterochronic switch gene from the study of a genetic screen for mutations that suppress the synthetic sterile phenotype of strain bearing LIN14 (n179) and egl-35 (n694) in *Caenorhabditis elegans* (Reinhart et al., 2000). The adult mutants die by bursting their vulva, therefore named lethal-7 (*Let-7*). Following identification of lin-4 as the first small temporal RNA (stRNA), *Let-7* next turned out to encode stRNA. Suppression of target mRNA by stRNAs was believed to be specific event to *C. elegans*. However, it was also found that the *Let-7* sequence is well conserved across species and *Let-7* is expressed in various human tissues (Pasquinelli et al., 2000). Furthermore, the stRNA synthesis mechanism was also common across species, which was discovered from studies on Dicer (an RNA-processing enzyme containing RNase III activity) knockout (KO) in *Drosophila melanogaster* and targeted disruption of Dicer mRNA in human cultured cell line, HeLa cells. These observations showed that Dicer plays a central role in stRNA generation as like in siRNA synthesis (Hutvágner et al., 2001). Nowadays, *Let-7* is known as a miRNA.

MiRNAs are encoded on genomes, and transcribed by RNA polymerase II and III like long RNAs and processed by Drosha (an RNA-processing enzyme containing RNase III activity) and DGCR8 (known as Pasha) in the nucleus (Lee et al., 2003; Han et al., 2004, Wang et al., 2007; Carthew and Sontheimer, 2009). The pre-miRNAs are further cleaved by Dicer to generate mature miRNAs in the cytoplasm. MiRNAs are incorporated into a multi protein complex, known as the RNA induced silencing complex (RISC). The RISC post-transcriptionally regulates mRNA expression by translational inhibition and/or degradation of target mRNAs through the interaction between target mRNAs and its miRNA (Carthew and Sontheimer, 2009). The base paring much between a stretch of 6-8 nucleotides at the 5’ end of miRNA, called a “seed” region, and 3’UTR of mRNA is important for targeting of mRNA by miRNA (Lewis et al., 2005; Grimson et al., 2007). It has been shown that miRNAs which have similar seed sequence target same genes. MiRNAs which contain a similar seed region are categorized as family miRNA. The human *Let-7* miRNA families consist of 13 discrete genetic loci that resolve into 10 unique mature miRNA sequences, *Let-7*a, *Let-7*b, *Let-7*c, *Let-7*d, *Let-7*e, *Let-7*f, *Let-7*g, *Let-7*i, miR-98, and miR-202. Only the human miR-202, not the other organisms, is classified as a *Let-7* family miRNA. The *C. elegans* genome encodes 9 copies of *Let-7* family miRNAs, *Let-7*, miR-48, miR-84, miR-241, miR-265, miR-793, miR-794, miR-795, and miR-1821. Among them, *Let-7*, miR-48, miR-84, and miR-241 are experimentally detectable *Let-7* family miRNAs (Rouse and Slack, 2008). Interestingly from *Let-7* functional analysis, developmental timing and differentiation have been focused not only in invertebrates, but also in vertebrates. Despite *Let-7* was identified as the second miRNA after lin-4, its high evolutional conservation through animal species from *C. elegans* to human led to findings that miRNAs are general regulators of gene expression in various species. Here we review advanced research in understanding the contribution of *Let-7* and *Let-7* family to development and differentiation in invertebrates and vertebrates.

3. Roles of *Let-7* in Development

1) Invertebrates

I) Let-7 and Let-7 Family MiRNA in Developmental Timing in C. Elegans

(A) Discovery and Function of *Let-7*

C. elegans has been serving as a favorable model organism to study cellular and molecular basis underlying various developmental events, because *C. elegans* has a limited number of cells and is genetically well defined. The life cycle of *C. elegans* is 3.5 days (at 20 °C), it develops from egg to adult though four larval stages (L1-L4). By microscopic observation, cell linage in *C. elegans* is completely understood. Therefore, *C. elegans* is an excellent model to study the mechanisms of heterochrony in multicellular organisms (Sulston and Horvitz, 1977). Thus, it is not surprising that lin-4 and *Let-7*, the two founding members of the miRNA family were first identified in genetical analysis of genes regulating developmental timing, or heterochrony, in *C. elegans* (Reinhart et al., 2000; Lee et al., 1993; Moss et al., 2007). *Let-7* was initially found as a stRNA, following the identification of the first found stRNA, lin-4. At the end of larval development *Let-7* is dramatically upregulated with a significant reduction of key heterochronic proteins that promote larval-specific cell fates, thus ensuring the developmental transition into adulthood (Rougvie, 2005; Moss, 2007). In post-hatched larval stages, L1 to L4, hypodermal blast cells, known as seam cells, undergo cell division at each larval stage, and then exit from cell-cycle and differentiate, accompanying the expression of LIN28 protein, a zinc-finger transcription factor, at L4 to adult stages (Ambros and Horvitz, 1984; Rougvie and Ambros, 1995). *Let-7* mutants exhibit heterochronic abnormalities in the seam cell linage. Partial defect of *Let-7* activity results in retarded phenotypes. Their seam cells repeat L4 to adult cell division pattern without cell-cycle exit and undergo abnormal L5 to adult cell-cycle, then the cells exit from cell-cycle, showing that the defect of *Let-7* function retards progression from the juvenile state (Figure 1). In contrast, more complete defect of activity results in lethality by bursting through the vulva at the larval-to-adult transition (Reinhart et al., 2000). These abnormalities seem to arise from overexpression of *Let-7* target genes because it can be partially rescued by knocking down individual *Let-7* target genes (Slack et al., 2000; Abrahante et al., 2003). Further, a *Let-7* overexpression study showed increased dosage of *Let-7* promotes precociously cell-cycle exit and terminal differentiation of seam cells after L3 to L4 cell division, indicating that precocious *Let-7* expression results in premature differentiation of adult cell fates (Figure 1).

(B) *Let-7* Family MiRNA

In *C. elegans* there are other miRNAs which contain similar seed sequences with *Let-7*, including miR-48, miR-84, miR-241, miR-265, miR-793, miR-794, miR-795, and miR1821 (Rough and Slack, 2008). There has been no report about expression patterns and functions of miR-793, miR-794, miR-795, and miR-1821. Similar to *Let-7*, miR-48, miR-84, and miR-241 functions as heterochronic genes to regulate temporal patterning at the transition from L2 to L3 stage (Abbott et al., 2005).

While expression of *Let-7* RNA is detected at about L3 stage, the *Let-7* family miRNAs, miR-48 and miR-241 are observed at about L2 stage, and miR-84 appears from L1, the maximal expression of all *Let-7* family members are L4 (Esquela-Kerscher et al., 2005). Interestingly, single mutants of miR-48 and double mutants of miR-48 and miR-84 undergo extra adult molt. Additionally, miR-48 and miR-241 double mutants and miR-48, miR-241, and miR-84 triple mutants show reiteration of the L2 stage. Further, precocious accumulation of miR-48 accelerates developmental timing of seam cells and the vulva (Abbott et al., 2005; Li et al., 2005). These observations indicate that developmental timing and cell division required *Let-7* family miRNA function (Figure 1).

The loss of function (lf) mutant of *Let-7* causes a reiteration of the fourth larval (L4) cell fate at the adult stage. The gain of function (gf) mutant of *Let-7* exhibits precociously cell-cycle exit and terminal differentiation after the L3 to L4 molt. The miR-84 (lf) or miR-241 (lf) single mutants have no abnormality, but miR-48 (lf) single mutants cultured at 15°C and the miR-48 (lf); miR-84 double mutants exhibit supernumerary adult molt. The miR-48 (lf); miR241 (lf) double mutants, or the miR-48 (lf); miR-84 (lf); miR-241 (lf) triple mutants promote reiteration of the L2-prolifalative stage. The lin-4 (lf) mutants repeat the L1 cell division pattern. The LIN41 (lf) mutants exhibit precocious cell-cycle exit and maturation. The LIN28 (lf) mutants skip the L2 division pattern. TD, terminal differentiation. Circled arrows indicate repeating the same cell division pattern.

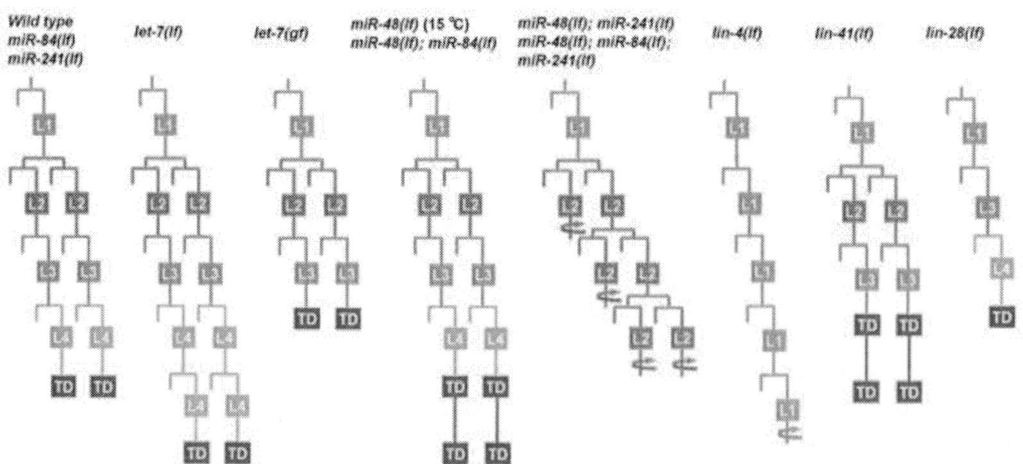

Figure 1. Schematic diagrams of cell lineage in *Let-7*-related *C. elegans* mutants of V1 to V4, and V6 seam cells.

(C) Regulation and Targets of *Let-7* MiRNA

By combination searching for base pairing match between *Let-7* and 3'UTR of heterochronic gene mRNAs and the genetic interaction analysis between *Let-7* and each heterochronic gene mutant, *Let-7* target mRNAs were identified to be LIN41, LIN14, LIN28, LIN42, and DAF12 (Reinhart et al., 2000). In fact, *Let-7* un-regulation at L4 stage down-regulates LIN41 gene, leading to up-regulation of LIN28 transcription factor. The conserved nuclear hormone receptors NHR23 and NHR25 regulate molting from L4 stage to adult. A *Let-7* paralog miR-84 acts synergistically with *Let-7* to promote terminal differentiation of the hypodermis and the cessation of molting in *C. elegans* through regulating NHR23 and NHR25 (Hayes et al., 2006).

Regarding biogenesis of miRNA, it has been demonstrated that RNA Polymerase II (Pol II) is primarily responsible for producing miRNA, while it should be noted that mature miRNAs can be produced from transgenic constructs driven by Pol II and Pol III-specific promoters (Lee et al., 2004; Zhou et al., 2005; Borchert et al., 2006; Zhou et al., 2008). Expression surge of primary *Let-7* (pri-*Let-7*) at L3 during development is critical for *Let-7* function as a heterochronic gene product. However, temporal regulation of *Let-7* transcription is one of the important issues that are still poorly understood. Regulatory region of *C. elegans Let-7* was investigated by transgenic analysis using an upstream sequence of *Let-7* miRNA, containing the putative promoter region, fused with gfp (Johnson et al., 2003). This sequence drove GFP expression to recapitulate the *Let-7* expression during *C. elegans* development. In this *Let-7* promoter sequence a short inverted repeat sequence was shown to be necessary and sufficient to drive *Let-7* expression pattern, however, binding transcription factors remain elusive. Genetic analysis showed that a nuclear receptor DAF12, functions immediately upstream of *Let-7*. However, DAF12 appears not to bind directly to the *Let-7* promoter and thus DAF12 regulation on *Let-7* is considered indirect. Recently, DAF12 with its steroid ligand was shown to directly activate promoters of miR-84 and miR-241, resulting in down-regulation of their target, a transcription factor hunch-back-like-1 (HBL1), allowing L2 to L3 transition (Bethke et al., 2009). DAF12 itself is a target of *Let-7* at later stages, suggesting that feedback loops play a role to regulate stage transitions (Grosshans et al., 2005). Notably, it was recently reported that HBL1 is responsible for inhibiting the transcription of *Let-7* in specific tissues until the L3, indicating that one important function of HBL1 in maintaining larval stage fates is inhibition of *Let-7* (Rouse and Slack, 2009). Thus, this study revealed *Let-7* as the target of the HBL1 transcription factor in *C. elegans* and suggest that a negative feedback loop mechanism for *Let-7* and HBL1 regulation constitute an important machinery to regulate *Let-7* temporal expression. The conserved nuclear hormone receptors NHR23 and NHR25 regulate molting from L4 stage to adult. A *Let-7* paralog miR-84 acts synergistically with *Let-7* to promote terminal differentiation of the hypodermis and the cessation of molting in *C. elegans* through regulating NHR23 and NHR25 (Hayes et al., 2006) (Figure 2).

In addition, interestingly it was found that let-60, encoding the *C. elegans* orthologous of the human oncogene RAS, is a direct target of *Let-7* and miR-84 miRNA *in vivo* (Johnson et al., 2005). Messenger RNAs of three types of human RAS family members, HRAS, KRAS, and NRAS contain multiple *Let-7* family miRNA targeting sites (Johnson et al. 2005). Although daf12 is a *C. elegans*-specific target of *Let-7*, other *Let-7* targets including LIN28, let-60 (RAS), and LIN41 (TRIM71) are conserved through evolution.

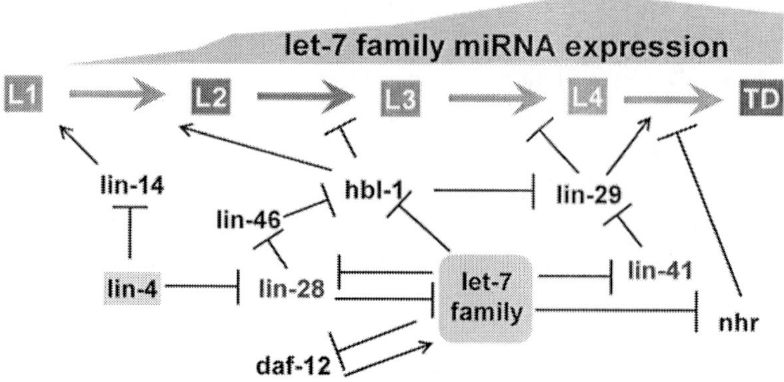

Figure 2. A model of heterochronic gene regulation regarding *Let-7* family miRNAs.

Let-7 family miRNAs are expressed from early L1 to terminal differentiated seam cells. The peak of *Let-7* expression is L4. After the L1 stage, individual seam cell lineage is regulated by *Let-7* family miRNAs through their target genes. Lin14 is a target of lin-4 at L1. Lin-4 is another heterochronic miRNA which expressed in L1 stage seam cells. Up-regulation of *Let-7* in the L4 results in down-regulation of LIN41, and consequently up-regulation of the transcription factor LIN28. Lin-46 functions immediately downstream of LIN28, affecting both the regulation of the heterochronic gene pathway. LIN28 and Lin41 are predicted to be important target genes also in vertebrates as well. HBL1 is responsible for inhibiting the transcription of *Let-7* in specific tissues until the L3. DAF12 is a *C. elegans*-specific target of *Let-7*.

(D) Modulation of *Let-7* Function

In addition to temporal regulators of *Let-7* co-factor is another regulator of *Let-7* function. A genetic modifier screen in *C. elegans* identified a ribosomal protein RPS14 as a cofactor of *Let-7* (Chan and Slack, 2009). RPS14 is able to modulate *Let-7* function in *C. elegans*. Reduction of RPS14 gene expression by RNAi suppressed the aberrant vulva and hypodermis development phenotypes of *Let-7* mutant animals and the mis-regulation of a reporter bearing the Lin41 3'UTR, a well established *Let-7* target. Thus, reduction of RPS14 activity leads to the elevation of *Let-7* function. The RPS14 protein co-immunoprecipitated with the nematode Argonaute homolog, ALG1, suggesting that RPS14, may interact with RISC and play roles in miRNA-mediated translational repression.

II) Functions of Let-7 in Drosophila Melanogaster

(A) Expression and Function of *Let-7*

Let-7 gene is not restricted to *C. elegans* but rather evolutionarily conserved throughout animal species up to human (Pasquinelli et al., 2000). In *Drosophila*, *Let-7* also plays an important role in controlling the juvenile-to-adult transition (Caygill and Johnston, 2008; Sokol et al., 2008). Although the potential family miRNAs of *Let-7*, miR-48 and miR-84, in *Drosophila* have been predicted to exist (Lau et al., 2001) and these miRNAs have been undetectable by Northern blots of RNAs from late third instar larvae through pupae (Bashirullah et al., 2003). Similar to *C. elegans Let-7*, it was observed that the expression of

Drosophila Let-7 also begins during the larval-to-adult transition. *Drosophila Let-7* expression is first detected about four hours before puparium formation, the same timing to the end of the third larval instar, and then highly accumulated during pupal development (Hutvágner et al., 2001; Bashirullah et al., 2003; Sempere et al., 2002). The *Drosophila Let-7* is processed from a common RNA precursor that contains two other conserved miRNAs, miR-100 and miR-125 (lin-4) (Bashirullah et al., 2003; Sempere et al., 2002; Sempere et al., 2003). All of these three miRNAs, called as *Let-7*-complex (*Let-7*-C), are coexpressed spatially and temporally. Two research groups investigated the function of *Drosophila Let-7* by generating *Let-7*-C knock out flies. Caygill and Johnston (2008) reported that *Drosophila* mutant lacking *Let-7* and miR-125 led to exhibit widespread phenotypes arising during metamorphosis. The mutant flies showed delay in maturation of abdominal neuromuscular junctions (NMJs). They attributed this phenotype to persistent expression of Abrupt, a *Let-7* target gene, in mutant muscle cells. In addition, they demonstrated that *Let-7* is critical for the appropriate timing of cell-cycle exit of developing wing cells, leading to a small wing phenotype. Sokol et al. (2008) observed that *Let-7*c transcripts are primarily expressed in the pupal and adult neuromusculature. The *Let-7*c mutant flies appear normal but displayed abnormal adult behavior such as severely reduced motility, flight, and fertility. They also found that *Let-7*c molecules are widely expressed in various tissues, including the central nervous system, motor neurons, and muscle cells. Consistent with this observation, the *Let-7*c mutant showed abnormal reorganization of neuromusculature. They found that the dorsal internal oblique muscles (DIOMs), which normally decay within 12 hrs of eclosion (Crossley, 1978; Kimura and Truman, 1990), remains in adult *Let-7*c mutants. It should be noted that they observed that CNS development occurs normally in *Let-7*c mutants.

Sokol et al. (2008) showed that *Let-7* miRNA alone is required for the *Let-7*c-dependent larval-to-adult remodeling of the abdominal neuromusculature through rescue experiment. Taken together, these two studies strongly suggested that *Let-7*c mutant phenotypes are heterochronic ones, because *Let-7*c mutant adults exhibited both juvenile characters such as immature neuromusculature as well as mature adults features such as normal external appearance.

(B) Targets of *Let-7* in *Drosophila Melanogaster*

From the study on *Let-7*c mutant flies, the roles of *Let-7*, as a larval-to-adult remodeling factor of the neuromuscular junction in the abdominal muscle and a cell-cycle exit inducer in the wing during metamorphosis were revealed. Caygill and Johnston (2008) found that the key function of *Let-7* is to suppress of Abrupt gene. Abrupt encodes a nuclear BTB-zinc finger regulatory protein and is broadly expressed during development and required for various functions, including regulating embryonic specificities of the neuromuscular connections, dendrite arborization in embryos (Hu et al., 1995; Li et al., 2004; Sugimura et al., 2004), and development of the fifth longitudinal vein of the wing (Cook et al., 2004). The 3'-untranslated region of the Abrupt mRNA contains at least five *Let-7*-binding sites and is regulated by *Let-7* expression in cultured cell assays (Burgler and Macdonald, 2005). Abrupt protein is expressed throughout the wind disc in late L3 flies (Cook et al., 2004), but its expression is lost early in the pupal stage (Caygill and Johnston, 2008). By using Gal4-mediated misexpression system *in vivo*, they demonstrated that misexpression of UAS-*Let-7* in L3 wing disc cells was sufficient to suppress Abrupt expression prematurely. Interestingly, Abrupt protein continues to be expressed in *Let-7*, miR-125 mutant in pupal stage wing discs.

They also showed that the persistent expression of Abrupt protein is due to loss of *Let-7* by using the Gal4-mediated misexpression system, indicating that *Let-7* is both necessary and sufficient for the appropriate temporal repression of Abrupt protein in wing discs. Furthermore, the delay in neuromuscular junction observed in *Let-7*, miR-125 mutant fly was suppressed by a partial loss of Abrupt function, showing that the persistent Abrupt expression in the abdominal muscles of *Let-7*, miR-124 mutants underlies the neuromuscular junction maturation defect. Thus, Abrupt was established as an important *in vivo* target of *Drosophila Let-7*. It should be noted that Abrupt is a target for *Drosophila Let-7*, but not a target for *Let-7* in *C. elegans* or mammals.

(C) Regulation of *Drosophila Let-7* Expression

In *Drosophila*, metamorphosis begins after the third instar larvae to pupae. The steroid hormone 20-hycroxyecdysone (ecdysone) regulates the metamorphosis timing from larval to adult stages. The *Drosophila Let-7* expression is induced in late third instar larvae in precise synchrony with ecdysone and ecdysone receptor (EcR) and highly expressed during pupal development (Pasquinelli et al., 2000; Hutvagner et al., 2001). The Broad-Complex (BR-C) gene functions at the top the hierarchy in the ecdysone pathway and plays an essential role to control expression of downstream target genes (Belyaeva et al., 1981; Zhimulev et al., 1982). BR-C encodes four isoforms of a zinc finger transcription factor. Sempere et al. (2003) examined whether *Let-7* expression is regulated by ecdysone in *Drosophila* using Northern blot analysis in mutant animal, organ cultures and S2 culture cells. They dissected late third instar larva to perform organ culture of imaginal discs, and incubated with or without ecdysone to measure *Let-7* expression. They found that ecdysone is actually required for *Let-7* expression during development. In addition, they also observed loss of *Let-7* expression in BR-C-null mutant animals, indicating that early ecdysone-inducible BR-C is required for *Let-7* expression during metamorphosis. These results showed that ecdysone signaling pathway is closely associated with the heterochronic pathway and implies that hormone-induced expression of *Let-7* plays a significant role in regulation of developmental stage transitions. There is a report that *Let-7* induction occurs independently of either ecdysone or EcR expression, suggesting that *Let-7* expression is not directory regulated by the ecdysone pathway (Bashirullah et al., 2003).

(D) Modulation of *Drosophila Let-7* Function

Parkinson's disease (PD) is a common neurodegenerative disorder and its main symptom is the selective and progressive loss of dopaminergic neurons. During the last decade, several loci were identified to be causative of familial Parkinson's disease. Mutations in Leucine-rich repeat kinase 2 (LRRK2) appear to be the most common genetic cause of a dominant form of PD (Paisán-Ruíz et al. 2004; Zimprich et al. 2004). Gain-of-function mutations of LRRK2 are linked with sporadic forms of PD as well as familial ones. Although molecular mechanism by which LRRK2 mutations cause PD has been unknown, it was recently reported that LRRK2 interacts with the miRNA pathway to regulate protein synthesis (Gehrke et al. 2010). In *Drosophila*, *Let-7* is expressed in dopaminergic neurons, and homozygous *Let-7* mutant showed decreased locomotor activity and a specific decrease in dopaminergic neurons. *Drosophila* e2f1 RNA is translationally repressed by *Let-7*. Gehrke et al. (2010) found that LRRK2 with pathogenic mutations suppressed *Let-7* function and interacts with RISC component Argonaute in *Drosophila*. E2F1 overexpression was sufficient to cause the degeneration of dopaminergic neurons and decreased locomotor activity. Conversely, increasing *Let-7* expression attenuated pathogenic LRRK2 effects. Thus, this study implies that the regulation of *Let-7* function by LRRK2 plays an important role in survival and maintenance of dopaminergic neurons.

2) Vertebrates

(I) Expression of Let-7 in Vertebrate Development

The human encodes 13 members, the mouse encodes 14 members, the chick encodes 11 members, Xenopus tropicalis encodes 9 members, and the zebrafish encodes 19 members of the *Let-7* family miRNAs. Among 13 human *Let-7* family members, *Let-7*a has identical sequence across various species from *C. elegans* to human. The increase of *Let-7* expression in chick limb development (Lancman et al., 2005; Kanamoto et al., 2006), mouse development (Lancman et al., 2005; Kanamoto et al., 2006), and neuronal differentiation of EC cells was reported (Schulman et al., 2005). Various studies showed that a major role of *Let-7* is to induce cells to exit cell-cycle and promote differentiation of cells. In mice, *Let-7* expression detected by Northern blot analysis increases after embryonic day 10.5 (E10.5) and peaks at E14.5, and then high level of its expression is maintained (Schulman et al., 2005). *Let-7* is undetectable in human and mouse embryonic stem cells, and *Let-7* expression increases upon differentiation (Wulczyn et al., 2007; Thompson et al., 2004). This *Let-7* expression is maintained in various adult tissues (Thompson et al., 2004; Sempere et al., 2004). Interestingly, the reduction of *Let-7* expression is observed in many human cancer cells (Park et al., 2007).

Regardless of several expression studies of *Let-7* in vertebrates, a definitive role of *Let-7* in development has remained elusive, probably due to redundant function by multiple *Let-7* family members in vertebrates. Existence of multiple family members makes *in vivo* direct experiments to draw a clear conclusion very difficult compared with those in invertebrates. Molecular mechanisms underlying the control of developmental timing in vertebrates are poorly understood, although many vertebrate's tissues are temporally regulated during development. Since mammals contain homologues of *Let-7*, a heterochronic gene in *C. elegans*, it has been hypothesized that similar heterochronic genes control developmental timing. Due to difficulty to analyze *Let-7* function *in vivo*, investigation of *Let-7* function in vertebrates has been mainly performed by using culture cell system as we will describe in the following section.

(II) A Possible Function of Vertebrate Let-7 during Development

Direct experiments to clarify *in vivo* function of *Let-7* in vertebrate development are practically very difficult as we described, however, there are couple of studies on *Let-7* target genes that may indirectly suggest a possible function of *Let-7* during vertebrate development. The *Let-7* miRNA regulates its targets by binding to complementary site in their 3'UTRs.

LIN28 was first identified in *C. elegans* as a heterochronic gene to regulate developmental timing (Moss et al., 2007; Ambros and Horvitz, 1984). LIN28 is an RNA-binding protein, containing a-cold-shock domain, and retroviral-type CCHC zinc fingers (Moss et al., 1997), and highly conserved through evolution. The mammalian homologs of

LIN28 are LIN28a and LIN28b, which bind to pre-*Let-7* and suppress production of the *Let-7* mature miRNA by inhibiting the pre-miRNA processing by Drosha or the loop cleavage reaction by Dicer (Heo et al., 2009). *Let-7* in turn directly suppresses LIN28 protein expression by binding to the LIN28 mRNA (Rybak et al., 2008; Nimmo and Slack, 2009). In order to investigate the function of LIN28/*Let-7* pathway *in vivo*, Zhu et al. (2010) generated transgenic mice to induce LIN28a expression. Upon LIN28a induction by doxycycline, the expressions of *Let-7*s were reduced in multiple tissues of LIN28-induced mice, and these mice showed increased body size and a proportional increase in organ sizes, and delayed onset of puberty. They found that LIN28a caused increased glucose utilization, a mechanism by which it may drive overgrowth *in vivo* by decreasing expression of *Let-7* family miRNA and increasing expression of *Let-7* target oncogenes such as MYC and RAS.

Another developmentally conserved *Let-7* target gene, Lin41, also known as TRIM71, encodes an RBCC-NHL protein, which has a RING, two B-boxes, and a coiled-coil domain associated with the Asn-His-Leu motif (Kanamoto et al., 2006). In the early embryonic stages of chick and mouse development, Lin41 is expressed in the developing limb buds, and the expression timing of *Let-7* and Lin41 is partially overlapped, suggesting that *Let-7* plays a role in limb development (Schulman et al., 2005). In *C. elegans*, it was proposed that LIN41 is temporally regulated by *Let-7* and lin-4 miRNAs that bind to complementary sites in the LIN41 3'UTR (Lancman et al., 2003). Lancman et al. (2003) reported that 3'UTR of chick Lin41 contains potential miRNA-binding sites for the chick orthologues of both *C. elegans Let-7* and lin-4, suggesting that these miRNAs regulate chicken Lin41. Interestingly, recent study reported that Lin41 functions as an E3 ubiquitin ligase for one of key components of RISC, Argonaute2 (AGO2), in stem cells, indicating the dual control mechanism for regulation of precocious *Let-7* function (Rybak et al., 2009). In mice, loss-of-function mutant of mouse Lin41 produced by gene-trap displayed neural tube closure defect and embryonic lethality, although their underlying mechanisms are unknown (Maller Schulman et al., 2008). As in *C. elegans* Lin41, mouse Lin41 seems to be regulated by *Let-7* and miR-125, a lin-4 homologue, through 3'UTR complementary sites for *Let-7* and miR-125. This fits well with the previously reported result that both human Lin41 and zebrafish LIN41 are regulated by *Let-7* (Lin et al., 2007).

The *Let-7* target sites in LIN41 3'UTR are also conserved in zebrafish. Misexpression of *Let-7* in zebrafish embryos resulted in severe embryonic developmental abnormality, suggesting that ectopic expression of *Let-7* causes precocious down-regulation of one or more endogenous *Let-7* targets (Kloosterman et al., 2004), suggesting that *Let-7* plays a significant role in zebrafish development. Lin et al. (2007) performed silencing experiment of the LIN41 orthologue by microinjection of RNA interference or morpholino in the zebrafish system and demonstrated developmental defects similar to the effects of *Let-7* misexpression experiment in zebrafish embryos reported by Kloosterman et al. (2008).

Fragile X syndrome (FXS) is the most common type of inherited cause of mental retardation, usually due to the loss of fragile X mental retardation protein 1 (FMRP1) encoded on the FMR1 gene. FMRP proteins link to the miRNA function through RISC component protein AGO. Fmrp1 is considered an important factor for synaptogenesis. Interestingly, knock down of *Let-7*c in rat hippocampal neurons lead to abnormal neuronal spine formation, and *Let-7*c is shown to associate with Fmrp1 in the mouse brain (Edbauer et al., 2010), however that function of *Let-7*c is still unknown.

Taken together, vertebrate *Let-7* may play an important functional role in various developmental events, including embryonic growth, limb development, and neural development and function.

4. Role of *Let-7* in Differentiation

1) *Let-7* in Stem Cell Pluripotency and Differentiation

The *Let-7* miRNA was found to function as a heterochronic gene in seam cell differentiation. Both expression and function of mature *Let-7* family miRNAs are controlled in stem cells through *Let-7* target genes. Neuronal induction by retinoic acid (RA) is commonly used for embryonic stem (ES) cell differentiation. The increased expression of mature *Let-7*a is detected in such RA-treated ES cells, embryonic neural stem cells, and the mouse E17 brain, but not in mouse ES cells in which pre-*Let-7*a-1 and pre-*Let-7*a-2 expression is detected (Rybak et al., 2009). Therefore, mature *Let-7* miRNA expression seems to be strictly controlled between at the embryonic stem cell stage and at its differentiation stage. One of the most important genes for regulation of *Let-7* family miRNAs, LIN28 has been known as one of pluripotency factors for induced pluripotent stem cells (iPS) (Yu et al., 2007; Okita et al., 2008), implying that down-regulation of *Let-7* is one of the key events for pluripotency maintenance.

DGCR8, DiGeorge syndrome critical region gene 8, encodes an RNA-binding protein required for miRNA biogenesis. The DGCR8-null mice exhibit loss of the production of canonical miRNAs (Han et al., 2004). Interestingly, DGCR8-null ES cells proliferate slowly and accumulate in G1 phase of its cell-cycle (Wang et al., 2007). The acceleration of the G1-S transition is essential for rapid cell-cycle of ES cells (White and Dalton, 2005). Overexpression of the ES cell-specific miRNAs, members of the miR-290 family, rescues the slow proliferation and G1 accumulation of the DGCR8-null ES cells. *Let-7* miRNAs seem to have an opposite role of the miR-294 family regarding cell-cycle maintenance through *Let-7* target genes, C-MYC, N-MYC, LIN28, SALl4, and NANOG (Melton et al., 2010). The C-MYC is one of the first iPS-inducible factors, and LIN28 is one of pluripotency factors for generating iPS cells from human somatic cells without usage of c-MYC (Yu et al., 2007; Okita et al., 2008). GFP (green fluorescence protein) expression in Oct4-gfp knock-in mouse embryonic fibroblasts (MEF) is a pluripotency monitor in iPS cell production. The number of GFP signal-expressing colonies in the OCT4, KIF2, and SOX2 infected MEF cells co-transfccted with *Let-7* inhibitor was significantly increased by 4.3 fold comparing with those in a control inhibitor-transfected cells, even without C-MYC co-expression (Melton et al., 2010). These findings indicate that down-regulation of *Let-7* family miRNAs in stem cell plays a critical role in acquirement and maintenance of pluripotency (Figure 3).

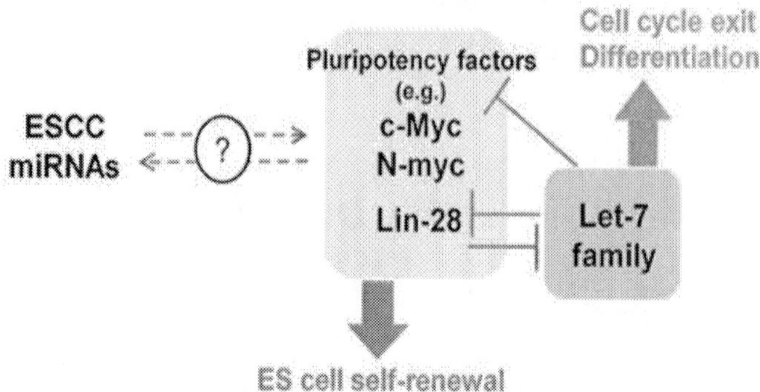

Figure 3. A model of the pluripotency maintenance by down-regulation of *Let-7* family miRNAs.

Embryonic stem cell cell-cycle (ESCC) miRNAs (e.g. miR-290 family) promotes expression of pluripotency factors. *Let-7* family miRNAs target pluripotency factor genes and lead ES cells to cell-cycle exit and differentiation. LIN28, one of the pluripotency factors and *Let-7* target genes, prevents mature *Let-7* family miRNA expression. Therefore, *Let-7* miRNAs play an opposite role against ESCC miRNAs in the ES cell pluripotency maintenance.

2) *Let-7* and Its Target Gene LIN28 in Neural Cell Fate Determination

The neural retina is developmentally a part of central nervous system (CNS) and known as a good model to analyze cell lineage in the CNS. In 1987, Tuner and Cepko (1987) demonstrated that a single neural progenitor cell in the rodent retina is able to generate multiple cell types including both neurons and glial cells. In the developing vertebrate retina, cone photoreceptor cells, ganglion cells, horizontal cells, and part of amacrine cells are generated from early progenitor cells. In contrast, rod photoreceptor cells, bipolar cells, amacrine cells, and Müller glial cells are generated mainly from late progenitor cells (Turner and Cepko, 1987). Similar to the retina, cortical progenitors generate glial cells after generating neurons (Pearson and Doe, 2004). As like ES cells, P19 embryonal carcinoma cells also possess a potential of differentiation along a neuronal-glial lineage by RA induction. Interestingly in these cells, constitutive expression of pluripotency factor LIN28 induces neurogenesis and inhibits gliogenesis (Blazer et al., 2010). In the developing mouse brain, the neurogenesis begins as early as from E10 and that reaches its peak around E12-14. Then after neurogenesis, gliogenesis begins around from E14 and a peak of gliogenesis comes around postnatal day 0 (P0) (Qian et al., 2000). In *C. elegans*, loss of function mutant of LIN28 skips the L2 program and shows premature entering to the L3 program (Figure 1). The gain of function mutant of LIN28 exhibits reiteration of L2 fates and delayed vulval development. These observations indicate that down-regulation of LIN28 in proper developing stages is necessary for proper cell differentiation, and the function of LIN28 in neuron-glia fate determination is well conserved across species.

However, it was recently found that *Let-7* family miRNAs are already expressed in the E9.5 mouse brain (Wulczyn et al., 2007). Furthermore, a recent report about miRNA expression by using deep sequencer indicated that in the E15.5 developing mouse brain *Let-7* family miRNAs are the most abundant miRNA more than miR-124, which is the most dominant miRNA in the adult brain (Ling et al., 2011), suggesting that *Let-7* is involved in neural cell fate determination rather than neuronal maturation. Additionally, it was reported that the stem cell regulator Tlx (Nr2e1), encoding an orphan nuclear receptor that is expressed in the vertebrate forebrain, and the cell-cycle regulator cyclin D1 are regulated by *Let-7* to control neural stem cell proliferation and differentiation (Zhao et al., 2010). Tlx maintains neural stem cells in an undifferentiated and self-renewable state to repress Tlx downstream target genes, p21 and Pten. Tlx and cyclin D1 are found to be a target of *Let-7*b. *Let-7*b inhibits Tlx and cyclin D1 expression by binding to the 3'UTR sites of their mRNAs. Overexpression of *Let-7*b in neural stem cells showed increased differentiation of both glial cells and neurons by suppressing Tlx expression. Moreover, *in utero* electroporation of *Let-7*b to embryonic neural stem cells resulted in reduced cell-cycle progression. These studies suggested that unknown molecular mechanisms, which are functionally associated between *Let-7* family miRNAs and LIN28, play a role in neuron and glia fate determination during vertebrate development.

3) Neural Regeneration and *Let-7*

When the vertebrate retina is injured, retinal Müller glial cells undergo dedifferentiation in the injured retinal region (Fisher and Reh. 2001). However, mechanisms underlying dedifferentiation of Müller glial cells are unknown. The capacity for regeneration of the injured retina seems to be strongest in fish, amphibians and chick. Therefore, zebrafish is one of good model systems to clarify the molecular mechanisms underlying Müller glial cell dedifferentiation. In the injured zebrafish retina, dedifferentiation of Müller glial cells begins 15 hours after injury, and then the Müller glial cells begin to proliferate two days after injury. The cell division becomes at its maximum 4 days after injury, and Müller glial cells stop their proliferation and begin differentiation (Fausett et al., 2001; Fausett et al., 2006). In those retinas, LIN28 expression is induced from at least 6 hours after injury and disappeared until 14 days after injury (Ramachandran et al., 2011). ASCL1A, achaete-scute complex homolog 1, encodes a member of the basic helix-loop-helix (bHLH) family of transcription factors. LIN28 transcription is directory activated by ASCL1A in Müller glial cells of the injured retinal region. The expression of *Let-7* miRNA in the Müller glial cells is down-regulated at least from 15 hours after injury. LIN28 inhibits mature *Let-7* miRNA expression. *Let-7* represses expression of regeneration-associated genes, including ASCL1A, HSPD1, LIN28, OCT4, PAX6B and C-MYC. In the uninjured retina, *Let-7* target genes, C-MYC and KIF4, and HSPD, are expressed in Müller glial cells. *Let-7* may inhibit this expression to prevent premature Müller glial cell dedifferentiation (Figure 4).

Notably, ASCL1A is a reciprocal target gene of *Let-7*, and a recent study shows that direct conversion of human fibroblasts to dopaminergic neuron requires Ascl1 expression (Pfisterer et al., 2011; Caiazzo et al., 2011). It may be interesting to examine if the downregulation of *Let-7* family miRNAs contributes not only to retinal neurogenesis in mammals but also to dopaminergic neurogenesis in the brain, which is related with etiology of Parkinson's disease.

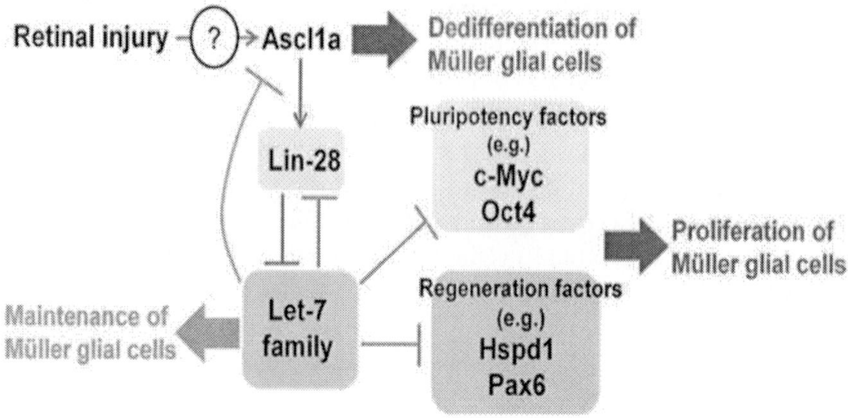

Figure 4. A model of the retinal regeneration by down-regulation of *Let-7* family miRNAs.

The expression of ASCL1A is simulated by retinal injury to repair retinal neurons and begins dedifferentiation of Müller glial cells. ASCL1A activates LIN28 expression and consecutive suppression of *Let-7* miRNA expressions occur in Müller glia cells. After repression of mature *Let-7*, both pluripotency factors and regeneration factors, which are targeted by *Let-7*, are upregulated at the protein level, and then Müller glial cells begin to proliferate. Therefore, the retinal regeneration is required for down-regulation of *Let-7* miRNAs.

Conclusion

A frequently-asked question is, "Do heterochronic genes exist in vertebrates, especially mammals?" A variation of this question is "Is mammalian *Let-7* homolog a developmental regulator?" Several recent studies implicated *Let-7* as a regulator of mammalian development. In *C. elegans* and *Drosophila melanogaster*, *Let-7* is a single locus and thus *Let-7* mutant provides us an important clue to reveal *Let-7* function. Studies using the *C. elegans* or *Drosophila* system designate *Let-7* as a heterochronic gene, although target genes of *Let-7* appear not to be same between these two species. On the other hand, the existence of multiple family members of *Let-7* in vertebrates hinders our investigation to clarify *in vivo Let-7* functions during development. At this point, we may still stand at a premature stage to conclusively place *Let-7* to a heterochronic gene in vertebrate development. The accumulation of further observation in the future *in vivo* studies is required to understand precise *Let-7* function in vertebrate development.

Despite *in vivo* loss of function studies in vertebrates are practically of extreme difficulty, functional study at cellular level has been more fruitful. It is of particular interest that mature *Let-7* is absent in ES cells or neural stem cells, however, *Let-7* expression is induced upon differentiation of those pluripotent cells. *Let-7* induces cell-cycle exit which is necessary for cell differentiation in multiple cell types. In accordance with this, it is also interesting that *Let-7* deregulation is observed in certain cancer cells. However, it is still unclear why there are so many *Let-7* family members in vertebrates and whether each of them has a different biological function and/or different target molecule during development. Our knowledge and understanding on *Let-7* in vertebrate development may be still in an emergence stage and further studies on *Let-7* may provide us not only an important clue to understand cell differentiation in development but also a useful tool for diagnosis and/or therapies in medical field.

REFERENCES

Abbott, A. L., Alvarez-Saavedra, E., Miska, E. A., Lau, N. C., Bartel, D. P., Horvitz, H. R. et al. The *Let-7* MiRNA family members miR-48, miR-84, and miR-241 function together to regulate developmental timing in Caenorhabditis elegans. *Dev. Cell.* 2005;9:403-414.

Abrahante, J. E., Daul, A. L., Li, M., Volk, M. L., Tennessen, J. M., Miller, E. A. et al. The *Caenorhabditis elegans* hunchback-like gene lin-57/HBL1 controls developmental time and is regulated by miRNAs. *Dev. Cell.* 2003;4:625-637.

Ambros, V., Horvitz, H. R. Heterochronic mutants of the nematode Caenorhabditis elegans. *Science.* 1984;226:409-416.

Balzer, E., Heine, C., Jiang, Q., Lee, V. M., Moss, E. G. LIN28 alters cell fate succession and acts independently of the *Let-7* miRNA during neurogliogenesis in vitro. *Development.* 2010;137:891-900.

Bashirullah, A., Pasquinelli, A. E., Kiger, A. A., Perrimon, N., Ruvkun, G., Thummel, C. S. Coordinate regulation of small temporal RNAs at the onset of *Drosophila* metamorphosis. *Dev. Biol.* 2003;259:1-8.

Belyaeva, E. S., Vlassova, I. E., Biyasheva, Z. M., Kakpakov, V. T., Richards, G., Zhimulev, I. F. Cytogenetic analysis of the 2B3-4-2B11 region of the X chromosome of Drosophila melanogaster. II. Changes in 20-OH ecdysone puffing caused by genetic defects of puff 2B5. *Chromosoma.* 1981;84:207-219.

Bethke, A., Fielenbach, N., Wang, Z., Mangelsdorf, D. J., Antebi, A. Nuclear hormone receptor regulation of miRNAs controls developmental progression. *Science.* 2009;324:95-98.

Borchert, G. M., Lanier, W., Davidson, B. L. RNA polymeRASe III transcribes human miRNAs. *Nat. Struct. Mol. Biol.* 2006;13:1097-1101.

Burgler, C., Macdonald, P. M. Prediction and verification of miRNA targets by MovingTargets, a highly adaptable prediction method. *BMC Genomics.* 2005;6:88.

Caiazzo, M., Dell'anno, M. T., Dvoretskova, E., Lazarevic, D., Taverna, S., Leo, D. et al. Direct generation of functional dopaminergic neurons from mouse and human fibroblasts. *Nature.* 2011;476:224-227.

Carthew, R. W., Sontheimer, E. J. Origins and Mechanisms of miRNAs and siRNAs. *Cell.* 2009;136:642-655.

Caygill, E. E., Johnston, L. A. Temporal regulation of metamorphic processes in *Drosophila* by the *Let-7* and miR-125 heterochronic miRNAs. *Curr. Biol.* 2008;18:943-950.

Cayouette, M., Poggi, L., Harris, W. A. Lineage in the vertebrate retina. *Trends Neurosci.* 2006;29:563-570.

Chan, S. P., Slack, F. J. Ribosomal protein RPS-14 modulates *Let-7* miRNA function in Caenorhabditis elegans. *Dev. Biol.* 2009;334:152-160.

Cook, O., Biehs, B., Bier, E. brinker and optomotor-blind act coordinately to initiate development of the L5 wing vein primordium in Drosophila. *Development.* 2004;131:2113-2124.

Crossley, A. C. The morphology and development of the Drosophila muscular system. Ashburner, M., Wright, T. R. F., editors. London: Academic Press; 1978.

Edbauer, D., Neilson, J. R., Foster, K. A., Wang, C. F., Seeburg, D. P., Batterton, M. N. et al. Regulation of synaptic structure and function by FMRP-associated miRNAs miR-125b and miR-132. *Neuron.* 2010;65:373-384.

Esquela-Kerscher, A., Johnson, S. M., Bai, L., Saito, K., Partridge, J., Reinert, K. L. et al. Post-embryonic expression of *C. elegans* miRNAs belonging to the lin-4 and *Let-7* families in the hypodermis and the reproductive system. *Dev. Dyn.* 2005;234:868-877.

Fausett, B. V., Goldman, D. A role for alpha1 tubulin-expressing Müller glia in regeneration of the injured zebrafish retina. *J. Neurosci.* 2006;26:6303-6313.

Fausett, B. V., Gumerson, J. D., Goldman, D. The proneural basic helix-loop-helix gene ASCL1A is required for retina regeneration. *J. Neurosci.* 2008;28:1109-1117.

Fischer, A. J., Reh, T. A. Müller glia are a potential source of neural regeneration in the postnatal chicken retina. *Nat. Neurosci.* 2001;4:247-252.

Gehrke, S., Imai, Y., Sokol, N., Lu, B. Pathogenic LRRK2 negatively regulates miRNA-mediated translational repression. *Nature.* 2010;466:637-641.

Grimson, A., Farh, K. K., Johnston, W. K., Garrett-Engele, P., Lim, L. P., Bartel, D. P. MiRNA targeting specificity in mammals: determinants beyond seed pairing. *Mol. Cell.* 2007;27:91-105.

Grosshans, H., Johnson, T., Reinert, K. L., Gerstein, M., Slack, F. J. The temporal patterning miRNA *Let-7* regulates several transcription factors at the larval to adult transition in C. elegans. *Dev. Cell.* 2005;8:321-330.

Han, J., Lee, Y., Yeom, K. H., Kim, Y. K., Jin, H., Kim, V. N. The Drosha-DGCR8 complex in primary miRNA processing. *Genes Dev.* 2004;18:3016-3027.

Hayes, G. D., Frand, A. R., Ruvkun, G. The miR-84 and *Let-7* paralogous miRNA genes of Caenorhabditis elegans direct the cessation of molting via the conserved nuclear hormone receptors NHR-23 and NHR-25. *Development.* 2006;133:4631-4641.

Heo, I., Joo, C., Kim, Y. K., Ha, M., Yoon, M. J., Cho, J. et al. TUT4 in concert with LIN28 suppresses miRNA biogenesis through pre-miRNA uridylation. *Cell.* 2009;138:696-708.

Hu, S., Fambrough, D., Atashi, J. R., Goodman, C. S., Crews, S. T. The *Drosophila* abrupt gene encodes a BTB-zinc finger regulatory protein that controls the specificity of neuromuscular connections. *Genes Dev.* 1995;9:2936-2948.

Hutvágner, G., McLachlan, J., Pasquinelli, A. E., Bálint, E., Tuschl, T., Zamore, P. D. A cellular function for the RNA-interference enzyme Dicer in the maturation of the *Let-7* small temporal RNA. *Science.* 2001;293:834-838.

Johnson, S. M., Grosshans, H., Shingara, J., Byrom, M., Jarvis, R., Cheng, A. et al. RAS is regulated by the *Let-7* miRNA family. *Cell.* 2005;120:635-647.

Johnson, S. M., Lin, S. Y., Slack, F. J. The time of appearance of the *C. elegans Let-7* miRNA is transcriptionally controlled utilizing a temporal regulatory element in its promoter. *Dev. Biol.* 2003;259:364-379.

Kanamoto, T., Terada, K., Yoshikawa, H., Furukawa, T. Cloning and regulation of the vertebrate homologue of LIN41 that functions as a heterochronic gene in Caenorhabditis elegans. *Dev. Dyn.* 2006;235:1142-1149.

Kimura, K. I., Truman, J. W. Postmetamorphic cell death in the nervous and muscular systems of Drosophila melanogaster. *J. Neurosci.* 1990;10:403-411.

Kloosterman, W. P., Wienholds, E., Ketting, R. F., Plasterk, R. H. Substrate requirements for *Let-7* function in the developing zebrafish embryo. *Nucleic Acids Res.* 2004;32:6284-6291.

Lancman, J. J., Caruccio, N. C., Harfe, B. D., Pasquinelli, A. E., Schageman, J. J., Pertsemlidis, A. et al. Analysis of the regulation of LIN41 during chick and mouse limb development. *Dev. Dyn.* 2005;234:948-960.

Lau, N. C., Lim, L. P., Weinstein, E. G., Bartel, D. P. An abundant class of tiny RNAs with probable regulatory roles in Caenorhabditis elegans. *Science.* 2001;294:858-862.

Lee, R. C., Feinbaum, R. L., Ambros, V. The *C. elegans* heterochronic gene lin-4 encodes small RNAs with antisense complementarity to LIN14. *Cell.* 1993;75:843-854.

Lee, Y., Ahn, C., Han, J., Choi, H., Kim, J., Yim, J. et al. The nuclear RNase III Drosha initiates miRNA processing. *Nature.* 2003;425:415-419.

Lee, Y., Kim, M., Han, J., Yeom, K. H., Lee, S., Baek, S. H. et al. MiRNA genes are transcribed by RNA polymeRASe II. *EMBO J.* 2004;23:4051-4060.

Lewis, B. P., Burge, C. B., Bartel, D. P. Conserved seed pairing, often flanked by adenosines, indicates that thousands of human genes are miRNA targets. *Cell.* 2005;120:15-20.

Li, M., Jones-Rhoades, M. W., Lau, N. C., Bartel, D. P., Rougvie, A. E. Regulatory mutations of miR-48, a *C. elegans Let-7* family MiRNA, cause developmental timing defects. *Dev. Cell.* 2005;9:415-422.

Li, W., Wang, F., Menut, L., Gao, F. B. BTB/POZ-zinc finger protein abrupt suppresses dendritic branching in a neuronal subtype-specific and dosage-dependent manner. *Neuron.* 2004;43:823-834.

Lin, Y. C., Hsieh, L. C., Kuo, M. W., Yu, J., Kuo, H. H., Lo, W. L. et al. Human TRIM71 and its nematode homologue are targets of *Let-7* miRNA and its zebrafish orthologue is essential for development. *Mol. Biol. Evol.* 2007;24:2525-2534.

Ling, K. H., Brautigan, P. J., Hahn, C. N., Daish, T., Rayner, J. R., Cheah, P. S. et al. Deep sequencing analysis of the developing mouse brain reveals a novel miRNA. *BMC Genomics.* 2011;12:176.

Maller Schulman, B. R., Liang, X., Stahlhut, C., DelConte, C., Stefani, G., Slack, F. J. The *Let-7* miRNA target gene, MLIN41/TRIM71 is required for mouse embryonic survival and neural tube closure. *Cell Cycle.* 2008;7:3935-3942.

Melton, C., Judson, R. L., Blelloch, R. Opposing miRNA families regulate self-renewal in mouse embryonic stem cells. *Nature.* 2010;463:621-626.

Moss, E. G., Lee, R. C., Ambros, V. The cold shock domain protein LIN28 controls developmental timing in *C. elegans* and is regulated by the lin-4 RNA. *Cell.* 1997;88:637-646.

Moss, E. G. Heterochronic genes and the nature of developmental time. *Curr. Biol.* 2007;17:R425-434.

Nimmo, R. A., Slack, F. J. An elegant miRror: miRNAs in stem cells, developmental timing and cancer. *Chromosoma.* 2009;118:405-418.

Okita, K., Nakagawa, M., Hyenjong, H., Ichisaka, T., Yamanaka, S. Generation of mouse induced pluripotent stem cells without viral vectors. *Science.* 2008;322:949-953.

Paisán-Ruíz, C., Jain, S., Evans, E. W., Gilks, W. P., Simón, J., Van der Brug, M. et al. Cloning of the gene containing mutations that cause PARK8-linked Parkinson's disease. *Neuron*. 2004;44:595-600.

Park, S. M., Shell, S., Radjabi, A. R., Schickel, R., Feig, C., Boyerinas, B. et al. *Let-7* prevents early cancer progression by suppressing expression of the embryonic gene HMGA2. *Cell Cycle*. 2007;6:2585-2590.

Pasquinelli, A. E., Reinhart, B. J., Slack, F., Martindale, M. Q., Kuroda, M. I., Maller, B. et al. Conservation of the sequence and temporal expression of *Let-7* heterochronic regulatory RNA. *Nature*. 2000;408:86-89.

Pasquinelli, A. E., Ruvkun, G. Control of developmental timing by miRNAs and their targets. *Annu. Rev. Cell Dev. Biol*. 2002;18:495-513.

Pearson, B. J., Doe, C. Q. Specification of temporal identity in the developing nervous system. *Annu. Rev. Cell Dev. Biol*. 2004;20:619-647.

Pfisterer, U., Kirkeby, A., Torper, O., Wood, J., Nelander, J., Dufour, A. et al. Direct conversion of human fibroblasts to dopaminergic neurons. *Proc. Natl. Acad. Sci.* U S A. 2011;108:10343-10348.

Qian, X., Shen, Q., Goderie, S. K., He, W., Capela, A., Davis, A. A. et al. Timing of CNS cell generation: a programmed sequence of neuron and glial cell production from isolated murine cortical stem cells. *Neuron*. 2000;28:69-80.

Ramachandran, R., Fausett, B. V., Goldman, D. ASCL1A regulates Müller glia dedifferentiation and retinal regeneration through a LIN28-dependent, *Let-7* miRNA signalling pathway. *Nat. Cell Biol*. 2011;12:1101-1107.

Reinhart, B. J., Slack, F. J., Basson, M., Pasquinelli, A. E., Bettinger, J. C., Rougvie, A. E. et al. The 21-nucleotide *Let-7* RNA regulates developmental timing in Caenorhabditis elegans. *Nature*. 2000;403:901-906.

Rougvie, A. E., Ambros, V. The heterochronic gene lin-29 encodes a zinc finger protein that controls a terminal differentiation event in Caenorhabditis elegans. *Development*. 1995;121:2491-2500.

Rougvie, A. E. Intrinsic and extrinsic regulators of developmental timing: from miRNAs to nutritional cues. *Development*. 2005;132:3787-3798.

Roush, S., Slack, F. J. The *Let-7* family of miRNAs. *Trends Cell Biol*. 2008;18:505-516.

Roush, S. F., Slack, F. J. Transcription of the *C. elegans Let-7* miRNA is temporally regulated by one of its targets, HBL1. *Dev. Biol*. 2009;334:523-534.

Rybak, A., Fuchs, H., Hadian, K., SmiRnova, L., Wulczyn, E. A., Michel, G. et al. The *Let-7* target gene mouse LIN41 is a stem cell specific E3 ubiquitin ligase for the miRNA pathway protein Ago2. *Nat. Cell Biol*. 2009;11:1411-1420.

Rybak, A., Fuchs, H., SmiRnova, L., Brandt, C., Pohl, E. E., Nitsch, R. et al. A feedback loop comprising LIN28 and *Let-7* controls pre-*Let-7* maturation during neural stem-cell commitment. *Nat. Cell Biol*. 2008;10:987-993.

Schulman, B. R., Esquela-Kerscher, A., Slack, F. J. Reciprocal expression of LIN41 and the miRNAs *Let-7* and miR-125 during mouse embryogenesis. *Dev. Dyn*. 2005;234:1046-1054.

Sempere, L. F., Dubrovsky, E. B., Dubrovskaya, V. A., Berger, E. M., Ambros, V. The expression of the *Let-7* small regulatory RNA is controlled by ecdysone during metamorphosis in Drosophila melanogaster. *Dev. Biol*. 2002;244:170-179.

Sempere, L. F., Freemantle, S., Pitha-Rowe, I., Moss, E., Dmitrovsky, E., Ambros, V. Expression profiling of mammalian miRNAs uncovers a subset of brain-expressed miRNAs with possible roles in murine and human neuronal differentiation. *Genome Biol.* 2004;5:R13.

Sempere, L. F., Sokol, N. S., Dubrovsky, E. B., Berger, E. M., Ambros, V. Temporal regulation of miRNA expression in *Drosophila* melanogaster mediated by hormonal signals and broad-Complex gene activity. *Dev. Biol.* 2003;259:9-18.

Slack, F. J., Basson, M., Liu, Z., Ambros, V., Horvitz, H. R., Ruvkun, G. The LIN41 RBCC gene acts in the *C. elegans* heterochronic pathway between the *Let-7* regulatory RNA and the LIN-29 transcription factor. *Mol. Cell.* 2000;5:659-669.

Sokol, N. S., Xu, P., Jan, Y. N., Ambros, V. Drosophila *Let-7* miRNA is required for remodeling of the neuromusculature during metamorphosis. *Genes Dev.* 2008;22:1591-1596.

Sugimura, K., Satoh, D., Estes, P., Crews, S., Uemura, T. Development of morphological diversity of dendrites in Drosophila by the BTB-zinc finger protein abrupt. *Neuron.* 2004;43:809-822.

Sulston, J. E., Horvitz, H. R. Post-embryonic cell lineages of the nematode, Caenorhabditis elegans. *Dev. Biol.* 1977;56:110-156.

Thomson, J. M., Newman, M., Parker, J. S., Morin-Kensicki, E. M., Wright, T., Hammond, S. M. Extensive post-transcriptional regulation of miRNAs and its implications for cancer. *Genes Dev.* 2006;20:2202-7.

Thomson, J. M., Parker, J., Perou, C. M., Hammond, S. M. A custom microarray platform for analysis of miRNA gene expression. *Nat. Methods.* 2004;1:47-53.

Turner, D. L., Cepko, C. L. A common progenitor for neurons and glia persists in rat retina late in development. *Nature.* 1987;328:131-136.

Wang, Y., Medvid, R., Melton, C., Jaenisch, R., Blelloch, R. DGCR8 is essential for miRNA biogenesis and silencing of embryonic stem cell self-renewal. *Nat. Genet.* 2007;39:380-385.

White, J., Dalton, S. Cell cycle control of embryonic stem cells. *Stem. Cell Rev.* 2005;1:131-138.

Wulczyn, F. G., SmiRnova, L., Rybak, A., Brandt, C., Kwidzinski, E., Ninnemann, O. et al. Post-transcriptional regulation of the *Let-7* miRNA during neural cell specification. *FASEB J.* 2007;21:415-26.

Yu, J., Vodyanik, M. A., Smuga-Otto, K., Antosiewicz-Bourget, J., Frane, J. L., Tian, S. et al. Induced pluripotent stem cell lines derived from human somatic cells. *Science.* 2007;318:1917-1920.

Zhao, C., Sun, G., Li, S., Lang, M. F., Yang, S., Li, W. et al. MiRNA *Let-7*b regulates neural stem cell proliferation and differentiation by targeting nuclear receptor TLX signaling. *Proc. Natl. Acad. Sci.* U S A. 2010;107:1876-1881.

Zhimulev, I. F., Vlassova, I. E., Belyaeva, E. S. Cytogenetic analysis of the 2B3-4--2B11 region of the X chromosome of Drosophila melanogaster. III. Puffing disturbance in salivary gland chromosomes of homozygotes for mutation l(1)pp1t10. *Chromosoma.* 1982;85:659-672.

Zhou, H., Huang, C., Xia, X. G. A tightly regulated Pol III promoter for synthesis of miRNA genes in tandem. *Biochim. Biophys. Acta.* 2008;1779:773-779.

Zhou, H., Xia, X. G., Xu, Z. An RNA polymeRASe II construct synthesizes short-hairpin RNA with a quantitative indicator and mediates highly efficient RNAi. *Nucleic Acids Res*. 2005;33:e62.

Zhu, H., Shah, S., Shyh-Chang, N., Shinoda, G., Einhorn, W. S., Viswanathan, S. R. et al. LIN28a transgenic mice manifest size and puberty phenotypes identified in human genetic association studies. *Nat. Genet.* 2010;42:626-630.

Zimprich, A., Biskup, S., Leitner, P., Lichtner, P., Farrer, M., Lincoln, S. et al. Mutations in LRRK2 cause autosomal-dominant parkinsonism with pleomorphic pathology. *Neuron.* 2004;44:601-607.

INDEX

A

access, 16, 34
acetylation, 48
acid, 6, 45, 101, 106, 135
acute myeloid leukemia, 36, 37, 38
acute promyelocytic leukemia, 6
AD, 8
adenocarcinoma, 37, 77, 81, 92, 99
adenoma, 97
adenosine, 17, 20
adenovirus, 108
adhesion, viii, 25, 27, 28, 29, 95, 102
adhesions, 28
adipocyte, 92, 103, 104, 105, 108
adipose, 69, 92, 97, 104, 108
adipose tissue, 69, 92, 104, 108
adult stem cells, 51, 53, 55, 58, 62
adulthood, 127
adults, 92, 131
age, 3, 78, 93, 128, 130, 136
aggressive behavior, 113
aging process, 3, 5
airway inflammation, 107
allele, 77
allergic asthma, 60
alters, 17, 68, 98, 139
amino, 17
amphibians, 137
anchorage, 71, 96, 122
angiogenesis, 60, 61, 69, 91, 104, 106, 108, 117, 120
angiotensin II, 72
antagonism, 50
antibody, 97
anti-cancer, viii, 25, 31, 82
anticancer drug, 92
antigen, 30
antisense, 1, 47, 49, 66, 99, 141
antisense oligonucleotides, 47, 99
antitumor, 63
apoptosis, viii, ix, 2, 20, 21, 23, 25, 26, 27, 28, 30, 31, 36, 37, 40, 41, 60, 61, 62, 63, 72, 76, 82, 86, 90, 98, 109, 115, 117, 118, 123
arrest, 29, 30, 90, 113
aspiration, 77
asthma, 71
atherosclerosis, 104
ATP, 15, 113, 118
autoantibodies, 93
autoimmunity, 92, 93, 108
awareness, 4

B

base, 16, 57, 101, 126, 129
base pair, 16, 57, 101, 129
Beijing, 89
benign, 46, 107
bioavailability, 100
bioinformatics, 6
biological processes, vii, 26, 28, 40, 90, 96, 101
biological responses, 103
biological roles, x, 125
biomarkers, 36, 79, 85, 107
blood, 3, 60, 61
body size, 61, 101, 134
body weight, 101
brain, 59, 71, 87, 94, 107, 134, 135, 136, 137, 141, 143
branching, 141
Brazil, 109
breast cancer, 8, 26, 31, 38, 40, 46, 55, 70, 83, 84, 91, 92, 98, 104, 105, 106, 108
breast carcinoma, 117

C

Caenorhabditis elegans, vii, viii, 1, 2, 8, 20, 21, 43, 53, 54, 55, 58, 71, 72, 75, 76, 86, 94, 106, 122, 126, 139, 140, 141, 142, 143
Cairo, 35, 36
cancer cells, ix, 4, 6, 8, 18, 31, 36, 47, 54, 62, 65, 73, 78, 79, 82, 85, 92, 96, 98, 99, 103, 106, 109, 113, 114, 116, 118, 119, 133, 138
cancer death, 100
cancer progression, 3, 6, 31, 55, 78, 81, 85, 91, 106, 142
cancer stem cells, viii, 4, 43, 49, 51, 63, 82
cancer XE "cancer" therapy, viii, 51, 57, 63, 64, 66, 67, 92, 102
cancerous cells, 27, 35
candidates, vii, 12, 13
carcinogenesis, 63, 91
carcinoma, 8, 14, 20, 21, 27, 28, 30, 37, 39, 41, 45, 77, 82, 96, 112, 114, 120, 136
cardiomyopathy, 105
cardiovascular disease, 97, 101
cardiovascular disease XE "cardiovascular disease" s, 97
cascades, 46
catalysis, 20, 103
CCND2, 29, 37, 113, 117, 120
CD95, 121
CDK inhibitor, 29
cDNA, 54
cell biology, 113
cell cycle, vii, viii, ix, 1, 5, 25, 27, 29, 30, 44, 47, 50, 61, 62, 71, 90, 99, 109, 113, 115, 117, 118, 119, 122
cell cycle checkpoints, viii, 25
cell death, 17, 30, 41, 72, 81, 86, 90, 91, 108, 113, 118, 123, 141
cell differentiation, x, 21, 36, 53, 68, 76, 118, 126, 135, 136, 138
cell division, viii, 3, 26, 29, 43, 47, 48, 99, 117, 127, 128, 137
cell fate, 58, 64, 67, 68, 127, 128, 137, 139
cell line, 13, 14, 17, 18, 19, 20, 22, 28, 34, 39, 45, 47, 49, 59, 62, 63, 69, 70, 80, 81, 82, 85, 92, 93, 97, 98, 99, 111, 112, 113, 114, 116, 117, 118, 126, 128, 130, 136, 143
cell lines, 13, 17, 18, 19, 20, 45, 47, 62, 80, 81, 82, 85, 92, 93, 97, 98, 99, 111, 112, 113, 118
cell movement, 83
cell signaling, 28, 38, 67
cell surface, 17
cellular metabolism, vii, 11
cellular signaling pathway, 26
central nervous system, 131, 136
central nervous system XE "nervous system" (CNS), 136
cervical cancer, 96, 114
challenges, ix, 89
chemical, 100
chemotherapeutic agent, 79, 98
chemotherapy, 3, 35, 76, 77, 78, 79, 80, 81, 82, 83, 92, 113, 118, 119
chicken, 134, 140
China, 25, 37, 57, 67, 89
cholangiocarcinoma, 118
cholesterol, 101
chromosomal abnormalities, 58
chromosome, 110
chronic lymphocytic leukemia, 3, 6
cirrhosis, 38
City, 78
classes, 47
classification, 110
cleavage, 15, 16, 18, 22, 26, 134
clinical application, viii, 57, 81
clinical trials, 65
closure, 134, 141
clusters, 91
CNS, 131, 136, 142
coding, vii, viii, ix, 1, 2, 5, 15, 17, 39, 40, 46, 54, 57, 64, 67, 72, 75, 90, 105, 115, 125
colitis, 93
colon, 3, 18, 20, 21, 26, 30, 36, 37, 39, 70, 76, 77, 84, 91, 95, 99, 102, 113, 118, 121, 122
colon cancer, 3, 18, 21, 36, 70, 77, 84, 102, 113, 118, 121, 122
colorectal cancer, 13, 22, 37, 39, 79, 85, 87, 102
complementarity, 12, 18, 141
complexity, 8, 119
compounds, 98
conflict, 63
Congress, iv
conservation, 43, 94, 96, 127
constituents, 28
cooperation, 31, 44
coronary artery disease, 104
correlation, 17, 19, 46, 47, 79, 96, 98, 111, 113, 115
correlations, 7, 78
cortex, 103
crown, 61
crystal structure, 98
CSCs, 4
CSD, 118
CT, 8
cues, 142
culture, 132, 133

cyclins, ix, 109
cytoplasm, 2, 12, 15, 33, 99, 126
cytosine, 16
cytotoxic agents, 80
cytotoxicity, 41

D

database, 26, 27, 32, 37, 47
deaths, 113
decay, 15, 45, 93, 131
defects, 44, 50, 64, 134, 141
degradation, 1, 2, 12, 16, 18, 22, 30, 57, 64, 90, 91, 101, 116, 126
dendrites, 143
deposition, 101
deregulation, viii, 7, 12, 25, 26, 32, 35, 38, 92, 94, 95, 104, 110, 111, 113, 114, 115, 118, 122, 138
derivatives, 48
detectable, 14, 15, 59, 126
detection, 83
diabetes, 93
diet, 101
discrimination, 100
discs, 131, 132
disease progression, 78
diseases, vii, ix, 11, 12, 26, 32, 36, 40, 83, 89, 97, 101, 102
disorder, 132
diversity, 17, 143
DNA, 29, 32, 34, 47, 48, 69, 84, 90, 110
DNA damage, 29
DNA repair, 29, 90
Document Template, 63, 58, 65, ate, 67
dopamine, 104
dopaminergic, 93, 132, 137, 139, 142
dosage, 127, 141
down-regulation, vii, ix, x, 6, 11, 22, 39, 44, 45, 51, 70, 82, 92, 93, 109, 111, 113, 114, 116, 117, 118, 125, 129, 130, 134, 135, 136, 138
Drosophila, ix, 5, 8, 44, 53, 55, 58, 59, 68, 72, 76, 94, 103, 107, 125, 126, 130, 131, 132, 138, 139, 140, 141, 142, 143
drug resistance, vii, ix, 75, 78, 81, 82, 89, 92, 98, 109, 114, 118
drug therapy, 118
drug treatment, 115
drugs, 31, 35, 92, 98, 102, 103, 118

E

E-cadherin, 38, 91, 102, 103, 105, 106
ECM, 28
editors, 140
effusion, 45
egg, 127
electroporation, 137
embryogenesis, 45, 55, 81, 94, 96, 107, 142
embryonic development, viii, x, 45, 57, 125, 134
embryonic stem cells, 3, 45, 48, 49, 50, 51, 53, 54, 55, 69, 70, 71, 73, 94, 122, 133, 141, 143
encoding, 12, 35, 49, 91, 129, 137
endocrine, 93
endonuclease, 12, 17
endothelial cells, 61, 91
energy, 92
environment, 22, 39, 63, 70
enzyme, 12, 17, 18, 19, 21, 33, 51, 103, 126, 140
enzymes, 18, 26, 49
epigenetic alterations, 58
epigenetic modification, 33
epigenetics, 22
epithelial cells, 5, 59, 80, 82, 91, 97, 106
epithelial ovarian cancer, 8, 64, 69, 78, 79, 85, 87, 95, 105, 108
epithelium, 80, 121
equilibrium, 48
erythroid cells, 70
erythropoietin, 51
esophageal cancer, 76, 113
estrogen, 83
etiology, 137
eukaryotic, 17
evidence, viii, x, 4, 7, 15, 27, 29, 35, 45, 47, 49, 53, 66, 70, 75, 76, 77, 80, 82, 83, 90, 93, 96, 98, 102, 125
evolution, 96, 129, 133
examinations, 83
exons, 12
exposure, 77, 80, 84

F

families, viii, ix, 25, 26, 50, 54, 60, 62, 70, 75, 125, 126, 140, 141
family members, vii, viii, ix, 3, 5, 11, 13, 17, 25, 26, 27, 28, 30, 32, 35, 45, 46, 52, 53, 57, 58, 59, 65, 78, 82, 89, 90, 94, 95, 96, 98, 102, 110, 118, 128, 129, 133, 138, 139
FAS, ix, 17, 20, 22, 23, 30, 109, 117, 118
fasting, 93
fat, 84, 101
feedback loop, viii, 13, 15, 16, 17, 18, 20, 23, 35, 37, 41, 43, 46, 51, 55, 64, 65, 66, 71, 72, 117, 129, 142
fertility, 44, 131

fibroblasts, 39, 66, 117, 119, 135, 137, 139, 142
fibroids, 5
fibrosis, ix, 7, 40, 89, 97, 107
fine tuning, viii, 12, 43, 119
fish, 137
flight, 44, 131
fluorescence, 135
Ford, 84, 120
forebrain, 137
formation, 5, 44, 60, 62, 96, 98, 100, 113, 115, 131, 134
formula, 100
foundations, 43
fragile site, 90, 103, 110, 119
functional analysis, 20, 85, 113, 126

G

ganglion, 136
gastrointestinal tract, 114
gene expression, vii, viii, ix, 1, 6, 8, 11, 20, 25, 26, 27, 34, 38, 41, 43, 48, 49, 52, 76, 78, 79, 83, 90, 94, 98, 101, 104, 107, 125, 127, 130, 143
gene promoter, 99
gene regulation, x, 12, 70, 116, 125, 130
gene silencing, 20, 70
gene therapy, 22, 63, 65, 66, 68, 71
genetic alteration, ix, 63, 109
genetic defect, 139
genetics, 43, 94
genome, 32, 44, 76, 91, 110, 126
genomic instability, 91
genomic regions, 96, 103, 110, 119
genomics, 84, 85, 102
genotype, 77
germ layer, 60
Germany, 75
gland, 51, 59
glia, 59, 60, 71, 136, 137, 138, 140, 142, 143
glial cells, 136, 137, 138
glioblastoma, 4, 13, 21, 80, 81, 84, 103, 117
Glioblastoma, 107
glucagon, 93
gluconeogenesis, 93
glucose, 93, 101, 103, 104, 134
glucose tolerance, 101
grants, 67
growth, viii, ix, 25, 31, 34, 36, 40, 46, 47, 51, 54, 62, 64, 69, 71, 75, 83, 84, 85, 86, 90, 96, 102, 105, 107, 108, 109, 110, 113, 116, 117, 119, 120, 122, 135
growth arrest, 90
growth factor, 31, 34, 51, 62, 64, 69, 85, 105, 107, 108, 110, 113, 116, 122
guidelines, 43

H

hair, 8, 72
half-life, 116
harbors, 16
HCC, 30, 82
head and neck cancer, 3, 8, 78, 82, 87, 114, 118, 123
hematopoietic system, 51
hepatitis, 38, 41
hepatocellular cancer, 63, 82
hepatocellular carcinoma, 30, 31, 36, 38, 39, 40, 41, 63, 82, 86
hepatocytes, 82
hepatoma, 82
heterogeneity, 38
hippocampus, 103
histology, 79
histone, 22, 46, 48, 49, 54, 115
HM, 6
homeostasis, 93, 101, 104, 119
hormone, 5, 6, 129, 132, 139, 140
host, 19, 98
hub, 52
human cancers, vii, ix, 3, 6, 7, 11, 26, 27, 28, 35, 46, 89, 94, 96, 98, 121
human genome, vii, 1
Hunter, 20
hybridization, 101
hypermethylation, 64
hyperplasia, 92, 113
hypertrophy, 60, 61, 72, 92
hypodermis, 129, 130, 140
hypothesis, 49, 64, 82

I

ID, 33
identification, 6, 37, 40, 53, 65, 120, 126, 127
identity, 59, 67, 69, 142
idiopathic, 7, 40
IFN, 17, 22, 23
IL-13, 60
immune activation, 63
immune disorders, 92
immune function, 92
immune regulation, 103
immune response, 6, 36, 66, 92, 107
immune system, 31, 92, 93, 108
immunocompromised, 82

immunoprecipitation, 15, 48, 97
in utero, 137
in vitro, ix, x, 3, 23, 37, 66, 68, 82, 83, 86, 96, 97, 99, 101, 109, 113, 120, 125, 139
in vivo, x, 18, 37, 53, 61, 65, 66, 80, 83, 91, 96, 97, 98, 100, 101, 102, 105, 113, 116, 120, 125, 129, 131, 133, 134, 138
independence, 68
indolent, 3
inducer, 131
induction, 45, 46, 66, 83, 90, 91, 92, 98, 107, 114, 119, 132, 134, 135, 136
INF, 17
infection, 6, 7, 36, 38
inflammation, ix, 21, 38, 66, 68, 69, 89, 91, 93, 97, 104, 107, 108
inflammatory disease, 92, 103
inflammatory responses, 66
inhibition, 12, 22, 31, 45, 51, 62, 64, 71, 81, 82, 90, 93, 97, 98, 99, 100, 101, 102, 107, 116, 126, 129
inhibitor, 16, 45, 51, 98, 107, 114, 117, 135
initiation, viii, 4, 22, 47, 66, 67, 75, 83
injury, 137, 138
inner ear, 8, 72
insects, vii
insulin, 64, 69, 85, 93, 101, 104, 105, 106, 113, 116, 122
insulin sensitivity, 101, 104
integrated circuits, 65
integration, 106
integrin, 28
integrins, 28
integrity, 30, 91, 104, 116
interference, 21, 38, 102, 119, 134, 140
interferon, 21, 82, 118
internal oblique, 131
intron, 16
introns, 12, 19
invertebrates, ix, 125, 127, 133
ionizing radiation, 80, 81, 98, 103
iris, 5
irradiation, 3, 76, 77, 80, 81, 83
isoflavone, 79
isolation, 3
isozyme, 93
Israel, 72
issues, 45, 58, 63, 100, 129

J

Japan, 89, 125
Jordan, 4, 8

K

Kaposi sarcoma, 45
karyotype, 37

L

larva, 132
larvae, 130, 132
larval development, 127
larval stages, 127
laryngeal cancer, 7
larynx, 114
lead, 5, 18, 53, 66, 97, 110, 119, 120, 134, 136
Leahy, 8, 122
leiomyoma, 3, 96, 106
lens, 5, 7, 8, 59, 70, 72
lesions, 37
leucine, 59
leukemia, 4, 26, 63, 98
life cycle, 127
ligand, 3, 83, 129
light, 116
liver, 3, 35, 36, 40, 47, 60, 61, 72, 87, 93, 96, 101
liver cancer, 3, 35, 36, 47, 96
localization, 16
loci, 76, 110, 126, 132
locomotor, 132
locus, 26, 36, 84, 95, 102, 138
longevity, 5
luciferase, 17, 47
lung cancer, 4, 13, 14, 20, 22, 23, 26, 28, 32, 34, 37, 41, 45, 46, 47, 55, 65, 71, 72, 77, 80, 81, 83, 84, 85, 86, 87, 92, 95, 96, 98, 99, 100, 102, 103, 104, 107, 108, 110, 111, 113, 115, 116, 120, 122, 123
lung function, 97
Luo, 41, 72, 123
lupus, 93
lymph, 78
lymph node, 78
lymphoma, 4, 7, 26, 30, 39, 40, 45, 54, 65, 86, 95, 105, 117, 122
lysine, 48

M

machinery, 11, 76, 92, 93, 126, 129
majority, ix, 98, 99, 119, 125
malignancy, 72, 96, 114
malignant cells, 118
malignant melanoma, 39, 71, 122
malignant mesothelioma, 54, 95, 104

malignant tumors, 46
mammalian cells, 47
mammals, 55, 58, 59, 94, 132, 133, 137, 138, 140
management, 85
manipulation, 61
mass, 81, 92, 93, 101, 106
materials, 68
matrix, 28
maturation process, 64
medical, x, 126, 139
medicine, ix, 48, 109
MEK, 31, 45
melanoma, 3, 28
mellitus, 101
memory, 61, 68
mental retardation, 134
mesenchymal stem cells, 8, 54, 69, 104, 108
messenger RNA, 105, 108
metabolic disorder, 101
metabolism, vii, 11, 19, 76, 90, 101, 117
metamorphosis, 8, 55, 59, 72, 107, 131, 132, 139, 142, 143
metastasis, vii, 1, 4, 31, 36, 37, 46, 53, 69, 77, 78, 91, 102, 103, 104, 105, 112, 113, 121
metastatic cancer, 91
methylation, 34, 48, 68, 69, 78
mice, 3, 7, 54, 59, 66, 70, 73, 82, 86, 93, 96, 101, 113, 133, 134, 135, 144
microinjection, 134
microRNA, vii, 1, 4, 5, 6, 7, 8, 9, 17, 20, 21, 22, 23, 32, 36, 37, 38, 39, 40, 41, 53, 54, 55, 60, 62, 68, 69, 70, 71, 72, 102, 103, 104, 105, 106, 107, 108, 110, 119, 120, 121, 122, 123
midbrain, 93, 104
migration, 21, 31, 105, 121
Ministry of Education, 89
mitogen, 31, 46
mitosis, 29
model system, 137
models, 63, 65, 81, 84, 101, 103, 113, 118, 120
modifications, 17, 38, 48, 100
molecules, vii, viii, ix, 1, 5, 11, 12, 13, 20, 28, 62, 65, 71, 75, 76, 81, 83, 89, 90, 99, 100, 101, 122, 131
morphogenesis, 103, 126
morphology, 76, 140
mortality, 91
motif, 134
motor neurons, 131
mRNA, vii, 1, 2, 4, 11, 12, 15, 18, 22, 26, 57, 62, 63, 77, 91, 92, 93, 99, 108, 110, 113, 114, 116, 117, 118, 122, 126, 131, 134
mRNAs, 5, 17, 26, 59, 90, 100, 110, 111, 126, 129, 137
multicellular organisms, 127
multipotent, 3
muscle mass, 101
muscles, 44, 131, 132
mutant, 15, 16, 77, 93, 126, 128, 129, 130, 131, 132, 134, 136, 138
mutation, 2, 19, 44, 100, 143
mutations, 2, 13, 90, 96, 116, 126, 132, 141, 142
myosin, 113

N

nasopharyngeal carcinoma, 23, 81, 87
nematode, 5, 7, 96, 130, 139, 141, 143
nervous system, 136, 142
networking, 7, 70
neural development, 135
neuroblastoma, 36, 40
neurogenesis, 136, 137
neurons, 60, 93, 104, 132, 134, 136, 137, 138, 139, 142, 143
neutral, 81, 86
NMR, 98
non coding RNAs, vii, 1
normal development, x, 58, 125
Northern blot, 130, 132, 133
nucleic acid, 6, 99, 100, 105
nucleotides, viii, 16, 75, 90, 100, 126
nucleus, 2, 12, 16, 33, 99, 126
null, 132, 135

O

obesity, 93, 101, 103, 104
obstacles, 67
OH, 139
oncogenes, ix, 2, 3, 5, 12, 13, 26, 45, 46, 47, 58, 71, 81, 87, 89, 90, 95, 96, 98, 109, 110, 111, 134
oncogenesis, 26, 90, 104, 111, 114
oncogenic targets, ix, 89
oncoproteins, 119
oral cancers, 32, 37, 84, 120
oral cavity, 77, 114
organ, 3, 76, 132, 134
organism, 126, 127
ovarian cancer, 4, 6, 13, 26, 37, 38, 39, 41, 45, 53, 71, 78, 79, 82, 84, 85, 87, 92, 95, 96, 99, 107, 108, 113, 118, 119, 120
ovaries, 68
overlap, viii, 5, 25, 26, 47
overweight, 92

oxidative stress, 93, 104

P

p16INK4A, 29
p53, 12, 14, 29, 90, 104, 105, 106, 107
paclitaxel, 62, 79, 82, 85, 118
pairing, 7, 16, 54, 106, 116, 121, 140, 141
pancreas, 101
pancreatic cancer, 4, 7, 13, 23, 31, 39, 41, 45, 55, 78, 79, 81, 83, 84, 85, 86
parallel, 17, 68
parkinsonism, 144
patents, 67
pathogenesis, 27, 32, 99, 114
pathology, 144
pathways, viii, ix, 12, 17, 26, 27, 29, 31, 38, 43, 47, 49, 53, 54, 58, 67, 69, 84, 90, 98, 104, 107, 109, 116, 119, 121
PCR, 37, 40, 41, 50, 103
penetrance, 44
peripheral blood, 61, 63, 72
peripheral blood mononuclear cell, 61, 63
peritoneal cavity, 100
pharmaceutical, 103
pharmacogenomics, 92
phenotype, 2, 13, 27, 35, 45, 51, 52, 53, 93, 97, 106, 126, 131
phenotypes, 63, 73, 127, 130, 131, 144
phosphorylation, 28, 63, 114, 117, 118
plants, vii
plasma levels, 101
plasmid, 17
plasticity, 48
platform, 8, 107, 143
platinum, 4, 79, 82
point mutation, 15
polymerase, 1, 2, 12, 19, 21, 32, 99, 126
polymorphism, 37, 77, 79, 84, 87, 92, 103, 106, 120
polymorphisms, 32, 36, 92, 116
polysaccharide, 92
population, 71, 98, 116
positive feedback, 46, 117
post-transcriptional level, vii, viii, 1, 13, 17, 20, 25, 35, 58, 90
post-transcriptional regulation, vii, ix, 11, 13, 14, 15, 18, 20, 55, 63, 64, 71, 104, 109, 111, 115, 117, 119, 143
precursor cells, 80
prevention, 45
primary anticancer agents, ix, 75
primary tumor, 14
principles, 106, 108
progenitor cells, 3, 51, 52, 53, 59, 61, 62, 64, 65, 66, 67, 69, 98, 104, 119, 136
progesterone, 4, 83, 87
prognosis, ix, 35, 36, 41, 64, 69, 78, 79, 83, 85, 86, 87, 101, 102, 105, 108, 109, 110, 115, 116, 118
proliferation, viii, 2, 3, 4, 5, 7, 9, 21, 22, 23, 28, 29, 30, 31, 32, 36, 37, 38, 39, 43, 45, 46, 47, 48, 51, 53, 54, 58, 60, 62, 63, 64, 69, 70, 73, 76, 81, 86, 87, 90, 93, 94, 95, 96, 104, 111, 113, 115, 116, 117, 119, 120, 121, 122, 126, 135, 137, 143
promoter, 4, 14, 18, 19, 20, 21, 34, 65, 85, 97, 106, 115, 129, 140, 143
propagation, 3
prostate cancer, 7, 17, 22, 26, 29, 37, 40, 91, 96, 113, 120, 122
protection, 3
protein kinases, 28, 29
protein synthesis, 40, 132
proteins, 12, 14, 17, 19, 26, 28, 30, 33, 34, 35, 49, 51, 62, 78, 95, 96, 99, 101, 104, 107, 111, 113, 115, 116, 117, 118, 120, 122, 127, 134
proteome, 67
proto-oncogene, 29, 47
PTEN, 14, 41, 69, 91, 108
puberty, 61, 73, 134, 144
pumps, 113
purification, 18

R

race, 55
radiation, 4, 6, 80, 81, 83, 84, 98, 102, 113
Radiation, 84
radiation therapy, 113
radiation XE "radiation" treatment, 4, 80
radiotherapy, ix, 75, 77, 82, 85
reactions, 100
reading, 96
receptors, 36, 83, 105, 129, 140
recognition, 4, 98
recurrence, 78, 114
redundancy, 121
regeneration, viii, 5, 7, 8, 57, 59, 70, 71, 72, 137, 138, 140, 142
regenerative medicine, 48, 59
regulatory controls, 45
reintroduction, 113
relevance, 3, 67
remission, 67
renal cell carcinoma, 23
repair, 84, 138
replication, 29, 47, 110

repression, 1, 2, 7, 8, 16, 18, 22, 30, 35, 36, 46, 53, 57, 68, 103, 105, 106, 108, 111, 113, 114, 118, 119, 122, 130, 132, 138, 140
repressor, viii, 25, 26, 61, 83
requirements, 141
researchers, 2, 12, 65
resection, 100
residues, 15, 17, 18, 19
resistance, 3, 4, 34, 35, 41, 63, 67, 77, 78, 79, 81, 82, 84, 86, 92, 98, 102, 104, 105, 106, 108, 113, 115, 118, 119, 123
resolution, 66
response, viii, ix, 4, 6, 15, 36, 66, 75, 76, 77, 79, 80, 81, 82, 83, 84, 87, 91, 98, 106, 109, 122
restoration, 27, 45, 64, 102
retardation, 134
reticulum, 93
retina, 59, 136, 137, 139, 140, 143
retinoblastoma, 13, 96
RH, 7
ribozymes, 99
risk, 32, 37, 77, 79, 84, 95, 101, 114, 116, 120
RNAi, 19, 68, 102, 103, 130, 144
RNAs, vii, ix, 1, 5, 11, 12, 17, 47, 57, 72, 90, 99, 100, 102, 105, 107, 125, 126, 129, 130, 139, 141
rules, 101

S

safety, 65
salivary gland, 143
secrete, 44
secretion, 93, 101, 106
seed, viii, 7, 18, 22, 25, 26, 44, 126, 128, 140, 141
selectivity, 101
senescence, 2, 3, 8
sensitivity, 20, 30, 37, 60, 61, 78, 80, 83, 92, 98
sensitization, 22
sequencing, 17, 18, 54, 68, 69, 72, 107, 141
serotonin, 105
serum, 36
shock, 14, 15, 118, 133, 141
showing, 17, 20, 46, 79, 81, 127, 132
side effects, 65
signal transduction, 26, 27, 31
signaling pathway, vii, viii, 25, 27, 29, 30, 31, 32, 35, 91, 110, 115, 118, 132
signalling, 53, 71, 103, 106, 122, 142
signals, 20, 64, 143
silkworm, 105
Singapore, 43
siRNA, 15, 18, 100, 102, 113, 117, 126
skeletal muscle, 5, 6
skin, 44, 53, 94
smooth muscle, 9
SNP, 22, 32, 33, 37, 84, 116, 120
solid tumors, 41, 69, 91, 113, 123
somatic cell, 19, 45, 51, 58, 61, 119, 123, 135, 143
SP, 6, 7, 109
species, viii, 5, 13, 25, 26, 45, 57, 75, 76, 90, 94, 126, 127, 130, 133, 136, 138
spectroscopy, 98
spindle, 47
spine, 134
Spring, 8, 54, 122
squamous cell, 4, 6, 54, 78, 84, 86, 95, 103, 114, 117, 120, 121
squamous cell carcinoma, 4, 6, 54, 78, 84, 86, 95, 103, 114, 117, 120, 121
stability, 18, 100, 101, 118
stabilization, 30, 39, 84, 119, 122
state, viii, 4, 30, 43, 45, 48, 49, 51, 57, 58, 61, 63, 64, 127, 137
states, 51, 66
statistics, 121
stem cell differentiation, viii, 43, 48, 53
stem cell lines, 123, 143
stem cells, viii, 3, 5, 45, 46, 48, 49, 50, 51, 53, 54, 55, 57, 58, 59, 61, 62, 63, 64, 68, 69, 72, 82, 134, 135, 137, 138, 141, 142
stemness, viii, 3, 4, 5, 48, 49, 50, 51, 57, 58, 61, 62, 64, 66, 67, 78
sterile, 2, 126
stimulation, 83, 117, 118
stomach, 76
storage, 92
stress, 2, 3, 6, 7, 76, 84, 93
stress response, 2, 7, 76
stromal cells, 102
structure, 3, 16, 49, 107, 140
substrate, 48
substrates, 20, 28, 78, 83
succession, 68, 139
Sun, 6, 7, 38, 41, 60, 61, 62, 72, 73, 93, 97, 107, 143
supplementation, 99
suppression, 31, 46, 51, 67, 91, 93, 96, 104, 138
surveillance, 58
survival, vii, 4, 11, 23, 26, 30, 32, 37, 38, 39, 41, 45, 47, 51, 54, 55, 58, 63, 64, 70, 76, 77, 79, 81, 82, 84, 85, 86, 92, 95, 98, 100, 102, 105, 107, 108, 111, 113, 114, 116, 118, 119, 120, 121, 122, 123, 133, 141
survival rate, 77
syndrome, 134, 135
synthesis, 20, 47, 101, 126, 143

T

T cell, 61, 68, 93, 103, 108
T cell XE "T cell" s, 61, 68, 93, 103, 108
T lymphocytes, 118
Taiwan, 1, 57, 67
tamoxifen, 92, 106
TAP, 113
temperature, 2
TGF, 4, 60, 104
therapeutic agents, 107
therapeutic approaches, vii, ix, 11, 89, 119
therapeutic targets, 2, 12, 71
therapeutics, ix, 41, 68, 72, 81, 86, 92, 103, 106, 108, 109, 123
therapy, ix, 58, 63, 65, 67, 75, 76, 78, 80, 83, 86, 102, 104, 108, 113, 123
thermal stability, 100
thyroid, 13, 110, 114, 116, 120, 122, 123
thyroid cancer, 13, 114, 116
tissue, viii, 3, 12, 25, 27, 30, 46, 47, 61, 78, 93, 95, 100, 103, 114
tissue homeostasis, 30, 47
TLR, 31, 92
TNF, 90, 92, 117, 121
TNF-α, 90, 92
toxic effect, 81, 102
toxicity, 65
traits, 36
transcription, vii, viii, 1, 2, 4, 12, 13, 14, 18, 20, 25, 26, 29, 33, 35, 45, 46, 48, 50, 51, 52, 64, 66, 85, 91, 103, 117, 118, 127, 129, 130, 132, 137, 140, 143
transcription factors, vii, 1, 2, 12, 14, 26, 33, 48, 50, 51, 64, 129, 137, 140
transcripts, 13, 16, 18, 20, 22, 32, 45, 50, 64, 65, 67, 99, 131
transfection, 14, 17, 47, 81, 91
transformation, 5, 21, 26, 38, 39, 54, 58, 59, 66, 69, 95, 96, 104, 105, 106, 111, 116, 119, 120, 121, 123
transforming growth factor, 4, 97
translation, 2, 8, 11, 13, 18, 26, 108, 111, 118, 122
translocation, 7
transport, 107
treatment, vii, ix, 1, 3, 6, 17, 35, 65, 66, 67, 75, 76, 77, 79, 81, 82, 83, 92, 98, 99, 101, 102, 107, 109, 117, 118, 119
triggers, 60, 66
triglycerides, 93
tumor cells, 6, 30, 65, 67, 102
tumor development, vii, 1, 26, 35, 39, 54, 69, 105, 121
tumor growth, 41, 46, 81, 84, 96, 103, 107, 113, 120
tumor invasion, 46, 91, 104, 121
tumor metastasis, 45, 82
tumor necrosis factor, 17, 90
tumor progression, 6, 8, 23, 76, 78, 86, 90, 95, 105, 118
tumorigenesis, viii, ix, 2, 35, 40, 53, 57, 58, 63, 64, 65, 68, 90, 94, 109, 111, 113, 114, 115, 117, 119
tumors, viii, ix, 4, 35, 46, 63, 64, 65, 67, 75, 76, 77, 78, 79, 81, 83, 86, 89, 91, 95, 96, 99, 100, 101, 102, 106, 107, 110, 111, 112, 113, 114, 117
type 2 diabetes, 92, 101
tyrosine, 28

U

ubiquitin, 71, 108, 134, 142
underlying mechanisms, 3, 134
untranslated regions, 57, 90
USA, 11, 67

V

validation, ix, 47, 109, 110, 118
variations, 32
vascular diseases, 91, 108
vascular endothelial growth factor (VEGF), 91
vector, 65, 99
vein, 91, 131, 140
vertebrates, ix, 94, 125, 127, 130, 133, 138
viral vectors, 141
viruses, vii
vulva, 44, 126, 127, 128, 130

W

walking, 37
web, 39
wild type, 16, 19, 50, 51
Wnt signaling, 7
worldwide, 100, 101, 113
worms, 43, 84, 94

X

X chromosome, 139, 143
xenografts, 49, 99

Z

zebrafish development, x, 125, 134

zinc, 14, 15, 91, 103, 118, 127, 131, 132, 133, 140, 141, 142, 143